Tina Yu, Lawrence Davis, Cem Baydar, Rajkumar Roy (Eds.)

Evolutionary Computation in Practice

Studies in Computational Intelligence, Volume 88

Editor-in-chief
Prof. Janusz Kacprzyk
Systems Research Institute
Polish Academy of Sciences
ul. Newelska 6
01-447 Warsaw
Poland
E-mail: kacprzyk@ibspan.waw.pl

Further volumes of this series can be found on our
homepage: springer.com

Vol. 67. Vassilis G. Kaburlasos and Gerhard X. Ritter (Eds.)
Computational Intelligence Based on Lattice Theory, 2007
ISBN 978-3-540-72686-9

Vol. 68. Cipriano Galindo, Juan-Antonio
Fernández-Madrigal and Javier Gonzalez
*A Multi-Hierarchical Symbolic Model of the Environment
for Improving Mobile Robot Operation*, 2007
ISBN 978-3-540-72688-3

Vol. 69. Falko Dressler and Iacopo Carreras (Eds.)
*Advances in Biologically Inspired Information Systems:
Models, Methods, and Tools*, 2007
ISBN 978-3-540-72692-0

Vol. 70. Javaan Singh Chahl, Lakhmi C. Jain,
Akiko Mizutani and Mika Sato-Ilic (Eds.)
Innovations in Intelligent Machines-1, 2007
ISBN 978-3-540-72695-1

Vol. 71. Norio Baba, Lakhmi C. Jain and Hisashi Handa
(Eds.)
*Advanced Intelligent Paradigms in Computer
Games*, 2007
ISBN 978-3-540-72704-0

Vol. 72. Raymond S.T. Lee and Vincenzo Loia (Eds.)
Computation Intelligence for Agent-based Systems, 2007
ISBN 978-3-540-73175-7

Vol. 73. Petra Perner (Ed.)
Case-Based Reasoning on Images and Signals, 2008
ISBN 978-3-540-73178-8

Vol. 74. Robert Schaefer
Foundation of Global Genetic Optimization, 2007
ISBN 978-3-540-73191-7

Vol. 75. Crina Grosan, Ajith Abraham and Hisao Ishibuchi
(Eds.)
Hybrid Evolutionary Algorithms, 2007
ISBN 978-3-540-73296-9

Vol. 76. Subhas Chandra Mukhopadhyay and Gourab Sen
Gupta (Eds.)
Autonomous Robots and Agents, 2007
ISBN 978-3-540-73423-9

Vol. 77. Barbara Hammer and Pascal Hitzler (Eds.)
Perspectives of Neural-Symbolic Integration, 2007
ISBN 978-3-540-73953-1

Vol. 78. Costin Badica and Marcin Paprzycki (Eds.)
Intelligent and Distributed Computing, 2008
ISBN 978-3-540-74929-5

Vol. 79. Xing Cai and T.-C. Jim Yeh (Eds.)
*Quantitative Information Fusion for Hydrological
Sciences*, 2008
ISBN 978-3-540-75383-4

Vol. 80. Joachim Diederich
Rule Extraction from Support Vector Machines, 2008
ISBN 978-3-540-75389-6

Vol. 81. K. Sridharan
Robotic Exploration and Landmark Determination, 2008
ISBN 978-3-540-75393-3

Vol. 82. Ajith Abraham, Crina Grosan and Witold
Pedrycz (Eds.)
Engineering Evolutionary Intelligent Systems, 2008
ISBN 978-3-540-75395-7

Vol. 83. Bhanu Prasad and S.R.M. Prasanna (Eds.)
*Speech, Audio, Image and Biomedical Signal Processing
using Neural Networks*, 2008
ISBN 978-3-540-75397-1

Vol. 84. Marek R. Ogiela and Ryszard Tadeusiewicz
*Modern Computational Intelligence Methods for the
Interpretation of Medical Images*, 2008
ISBN 978-3-540-75399-5

Vol. 85. Arpad Kelemen, Ajith Abraham and Yulan Liang
(Eds.)
Computational Intelligence in Medical Informatics, 2008
ISBN 978-3-540-75766-5

Vol. 86. Zbigniew Les and Mogdalena Les
Shape Understanding Systems, 2008
ISBN 978-3-540-75768-9

Vol. 87. Yuri Avramenko and Andrzej Kraslawski
Case Based Design, 2008
ISBN 978-3-540-75705-4

Vol. 88. Tina Yu, Lawrence Davis, Cem Baydar and
Rajkumar Roy (Eds.)
Evolutionary Computation in Practice, 2008
ISBN 978-3-540-75770-2

Tina Yu
Lawrence Davis
Cem Baydar
Rajkumar Roy
(Eds.)

Evolutionary Computation in Practice

With 133 Figures and 29 Tables

 Springer

Dr. Tina Yu
Associate Professor
Department of Computer Science
Memorial University of Newfoundland
St John's, NLA1B 3X5
Canada
tinayu@cs.mun.ca

Dr. Lawrence Davis
President
VGO Associates
36 Low Street
Newbury, MA 01951
USA
ddavis@vgoassociates.com

Dr. Cem Baydar
Director
Peppers & Rogers Group
Buyukdere Cad. Ozsezen Is Merkezi
No: 122 C Blok Kat 8
Esentepe, Istanbul
Turkey
cbaydar@1to1.com

Professor Rajkumar Roy
Head of Decision Engineering Centre
Cranfield University
Cranfield
Bedford
MK43 0AL
UK
r.roy@cranfield.ac.uk

ISBN 978-3-540-75770-2   e-ISBN 978-3-540-75771-9

Studies in Computational Intelligence ISSN 1860-949X

Library of Congress Control Number: 2007940149

Contents

Contributing Authors

Carl Anderson is currently working with Archimedes, Inc. In 1998, he was awarded the Philip Steinmetz Fellowship at the Santa Fe Institute and in 2002–2003 was the Anderson/Interface Visiting Assistant Professor in Natural Systems at the School of Industrial and Systems Engineering at Georgia Institute of Technology, Atlanta (eggandchips@gmail.com).

Ronald C. Averill is President and CEO of Red Cedar Technology, Inc., and Associate Professor of Mechanical Engineering at Michigan State University (r.averill@redcedartech.com).

Laura Barbulescu is a Project Scientist in the Robotics Institute at Carniege Mellon University (laurabar@andrew.cmu.edu).

Derek Bartram is currently working towards a Ph.D. from Rail Research UK at the University of Birmingham for his project entitled A Computational Intelligence Approach to Railway Track Intervention Planning.

Cem Baydar is Director at Peppers & Rogers Group. Prior to Peppers & Rogers Group, he was the Director of Analytical Solutions at comScore Inc., USA. Prior to comScore, he worked for Accenture as a manager for 5 years (cem.baydar@gmail.com).

Peter Bentley is a Senior Research Fellow at the Department of Computer Science, University College, London (p.bentley@cs.ucl.ac.uk).

Daphna Buchsbaum is an artificial intelligence researcher and developer at Icosystem Corporation in Cambridge MA. In Fall 2007 she will begin an interdisciplinary PhD program in computational psychology at the University of California, Berkeley.

Eric Bonabeau is Chief Scientist of Boston-based Icosystem Corporation. He sits on the advisory board of a number of Fortune 500 corporations. Prior to his current position, Eric was the CEO of Eurobios. He has been a research director with France Telecom R&D and the Interval Research Fellow at the Santa Fe Institute (eric@icosystem.com).

Michael Burrow is a senior research fellow at the University of Birmingham Railway Group where he is working on several highway and railway related projects.

Lawrence Davis is President of VGO Associates and Vice President of VGO Oil and Gas (ddavis@vgoassociates.com).

Erik D. Goodman is Vice President of Technology at Red Cedar Technology, Inc. and Professor of Electrical & Computer Engineering and Mechanical Engineering at Michigan State University (goodman@egr.msu.edu).

Charles Neely Harper is Director of National Supply and Pipeline Operations at Air Liquide Large Industries U.S. LP. (Charles.Harper@Airliquide.com).

Gregory Hornby is a Project Scientist with U.C. Santa Cruz at NASA Ames Research Center. He received his Ph.D. in Computer Science from Brandeis University in 2002. He was a visiting researcher at Sony's Digital Creatures Laboratory in 1998 (hornby@email.arc.nasa.gov).

Adele Howe is Professor of Computer Science at Colorado State University (howe@cs.colostate.edu).

Arthur K Kordon is a Research and Development Leader in the Modeling Group within the Engineering and Process Sciences Corporate R&D Organization of the Dow Chemical Company (akkordon@dow.com).

Steven Manos is with Optical Fibre Technologoy Centre, University of Sydney, Australia and the Centre for Computational Science, University College London (s.manos@ucl.ac.uk).

Stefan Menzel is Senior Scientist at the Honda Research Institute Europe GmbH, Germany (stefan.menzel@honda-ri.de).

Jeff Potter is with Icosystem Corporation in Cambridge MA. He has collaborated with research teams at the MIT Media Lab, UC Berkeley, UT Austin, and UVC.

Kazuhiro Saitou is Associate Professor of Mechanical Engineering at University of Michigan, Ann Arbor, MI, USA (kazu@umich.edu).

Bernhard Sendhoff is Chief Technology Officer at the Honda Research Europe GmbH, Germany (bs@honda-ri.de).

Ranny Sidhu is a Lead Engineer at Red Cedar Technology, Inc. (r.sidhu@redcedartech.com).

Andrew Sutton is a Ph.D. student at Colorado State University.

Shingo Takeuchi is a graduate student at the Department of Mechanical Engineering at University of Michigan (stakeuch@umich.edu).

Darrell Whitley is Professor and Chair of Computer Science at Colorado State University (whitley@cs.colostate.edu).

Dave Wilkinson is a geophysicist in the Seismic Analysis and Reservoir Property Estimation Team at Chevron Energy Technology Company.

Xin Yao is a Professor of Computer Science at the University of Birmingham, Birmingham, U.K. He is also the Director of the Centre of Excellence for Research in Computational Intelligence and Applications (CERCIA), a Distinguished Visiting Professor of the University of Science and Technology of China, Hefei, and a visiting professor of three other universities (X.Yao@cs.bham.ac.uk).

Tina Yu is Associate Professor of Computer Science at Memorial University of Newfoundland, Canada. Prior to her current position, she worked at Chevron Technology Company for 6 years and Bell Atlantic (NYNEX) Science and Technology for 6 years (tinayu@cs.mun.ca).

Preface

Evolutionary Computation in Practice (ECP) has been a special track at the Genetic and Evolutionary Computation Conference (GECCO) since 2003. This track is dedicated to the discussion of issues related to the practical application of EC-related technologies. During the past four years, members from industry, governmental agencies and other public sectors have participated in presentations and discussions describing how evolution-related technologies are being used to solve real-world problems. They have also engaged in intense dialogue on bridging academic training and real-world usage of EC together.

This book compiles papers from practitioners who have presented their work at ECP. These contributing chapters discuss various aspects of EC projects, including:

- Real-world application success stories;

- Real-world application lessons learned;

- Academic case studies of real-world applications;

- Technology transfer to solve real-world problems.

We would like to thank Janusz Kacprzyk for inviting us to edit this book for Springer's *Studies in Computational Intelligence Series*. During the one-year period of book preparation, Thomas Ditzinger and Heather King at Springer have been very supportive to our needs. Dino Oliva has proofread many chapters of the book, which helped relieve some of the stress. We also thank the distinguished individuals who wrote foreword and back quotes for the book. Finally, the support of SIGEVO to this book project is greatly appreciated.

TINA YU, LAWRENCE DAVIS, CEM BAYDAR, RAJKUMAR ROY

Foreword

Give Evolutionary Algorithms a chance! Put them to work! But do it smart.

This book demonstrates not only that Evolutionary Algorithms (EAs) are now a mature technology that can (and should) be applied to solve large complex real-world optimization problems, but also that the diffusion between cutting edge research and outstandingly efficient (i.e. billion-dollars-saving) applications can be very fast indeed: the most recent algorithmic advances can be quickly put to work in domains that a priori seemed rather far from any Computer Lab.

The main characteristic of EAs that makes this possible – and clearly appears in all the chapters of this book – is their *flexibility*. Flexibility to explore non-standard search spaces: many representations used in the works described here involved both discrete and continuous variables, many are variable-length representations, and twisting the problem so that the use of more classical optimization algorithms would dramatically reduce the space of possible solutions. Along the same line, approaches pertaining to recently proposed embryogenic representations can be found here, together with revisited older ideas of indirect representations that had been used in scheduling for 20 years. Flexibility to optimize highly irregular and/or very expensive fitness functions, using specifically tailored flavors of EAs: hierarchical algorithms, surrogate models. Flexibility to efficiently handle very different types of constraints, in the representation itself, as well as in the morphogenetic process – the possibly complex mapping between genotypes and phenotypes – or in the fitness itself.

Of course, this flexibility has a cost: because EAs offer so many possible ways to achieve the same goal, there are many choices to make, ranging from their setup to the choice of representation and variation operators (crossover, mutation, and the like) and the tedious task of parameter tuning (as no general method yet exists to fully automatize this process). This leads to a situation that can be described in a way that is familiar to EA practitioners: when applying EA to real-world problems, too, there are no free lunches! Success stories such as those described in this book can only be obtained thanks to the close cooperation of open-minded experts in the application domain and smart evolutionary algorithmicists.

But in the end, such collaboration will pay off, by allowing what seems more and more necessary today when it comes to automatize repetitive complex tasks, and eventually try to improve on human operators: the re-introduction of the human factor. This goes from choosing representations that leave room for creativity (as there is nothing called artificial creativity, there are only creative programmers!), to letting human selection replace "natural" selection, what is done in the interactive evolution framework, and to designing optimized procedures that will be adapted to the user/customer (e.g. taking into account, when optimizing a delivery system, that Joe and Louis like to have lunch at Martin's Place ...).

So, even if we will not unveil here the address of Martin's Place, you must read this book, whether you are an EA practitioner wishing to start working on challenging problems that you will not find described in any textbook (and eventually willing to earn a few dollars at the same time), or an engineer willing to hear true success stories involving colleagues (or competitors!).

Marc Schoenauer, Ph.D.
Editor in Chief
Evolutionary Computation Journal
August, 2007

Chapter 1

AN INTRODUCTION TO EVOLUTIONARY COMPUTATION IN PRACTICE

Tina Yu[1] and Lawrence Davis[2]
[1]*Memorial University of Newfoundland;* [2]*VGO Associates*

Deploying Evolutionary Computation (EC) solutions to real-world problems involves a wide spectrum of activities, ranging from framing the business problems and implementing the solutions to the final deployment of the solutions to the field. However, issues related to these activities are not commonly discussed in a typical EC course curriculum. Meanwhile, although the values of applied research are acknowledged by most EC technologists, the perception seems to be very narrow: success stories boost morale and high profile applications can help to secure funding for future research and can help to attract high caliber students. In this book, we compiled papers from practitioners of EC with the following two purposes:

- Demonstrating applied research is essential in order for EC to remain a viable science. By applying EC techniques to important unsolved/or poorly-solved real-world problems, we can validate a proposed EC method and/or identify its weakness that restricts its applicability.

- Providing information on transferring EC technology to real-world problem solving and successful deployment of the solutions to the field.

1. APPLIED RESEARCH

In Chapter 2, Shingo Takeuchi and Kazuhiro Satitou applied a *multi-objective genetic algorithm* (MOGA) to design products that require disassembly for recycling at the end of their lives, according to recently established environmental regulations. One example of such products is electronic computers, where some of the parts can be reused and others are destined for recycling or a landfill. Product design is normally confronted with multiple objectives and constraints, such as technical specifications and economic returns. Recently, MOGA has become

T. Yu and L. Davis: *An Introduction to Evolutionary Computation in Practice*, Studies in Computational Intelligence (SCI) **88**, 1–8 (2008)
www.springerlink.com

a well-accepted method to obtain a set of non-compromising (Pareto optimal) solutions within a reasonable time frame.

In their case study of the design of a disassemble-able Mac Power G4 computer, the solutions have to satisfy four objectives and four constraints. Among them, two are related to the environmental regulations: maximizing the profit from reusing the disassembled components and minimizing the environmental impacts when the disassembled components are treated. To handle this design task, they expanded their MOGA with a double-loop, where the disassembly sequence planning is embedded within the 3D computer layout design. This modified MOGA delivered 37 alternative designs that have different trade-offs. This indicates that the modified MOGA can be applied to design other products requiring a similar embedded disassembly capability.

Chapter 3 focuses on a different issue which occurs frequently in the structural design of large and complex systems: how to perform component design without carrying out the time-consuming computer simulation of the entire system. In other words, how do we decompose the system such that only component simulations are needed to achieve a good global design? This is a common issue in the structural design of motor vehicles. Based on their extensive experience in delivering EC solutions to automotive industry, Erik Goodman, Ronald Averill and Ranny Sidhu presented two methods to address the decomposition problem.

The first method, called Component Optimization within a System Environment, is for systems with tightly coupled components, such as an automotive chassis. The method decomposes a system into a hierarchy. The boundaries of each component are updated at the end of each component design step. With a small number of iterations, they showed that this method can achieve an improved global design.

The second method is for systems that are composed of quasi-independent components, such as a lower compartment rail. Each component design is carried out independently, yet the components share their design solutions with components at the lower level of the system hierarchy. Since the upper level components are coarse approximations of the lower level components, the information helps reduce the potential design space of the lower level components and allows an improved global design to be obtained in a reasonable time frame. Although the authors only reported their implementation results on automotive component design, the potential of these methods on other large system structural design problems is worth exploring.

The researchers at the Honda Research Institute Europe also face the similar issue of expensive computer simulation required for aircraft structure design. This makes it impractical to use a large number of parameters to represent the design objects. As a result, the flexibility of the structure design is compromised, i.e. not all possible structure shapes are represented. This shortcoming has motivated Stefan Menzel and Bernhard Sendhoff to adopt a new kind of representation, called deformation method, where the design is represented as a lattice of control points of the object. Modification of these control points would produce a new object shape and also generate the grid points of the computational mesh, which is needed for the computational fluid dynamic mesh during simulation. Thus, this representation not only increases the design flexibility but also reduces computational simulation cost.

In Chapter 4, two kinds of deformation methods were investigated: free-form deformations (FFD) and direct manipulation of free form deformations (DMFFD). In FFD, the control points of the object lattice are used to represent the design object. In DMFFD, object points that can be placed on the design shape are used to represent the design object. The object points are then mapped into control points using a least squared algorithm. The authors have used both representations in their evolutionary algorithm to design a turbine stator blade that is part of a gas turbine for a small business jet. Their results show that the evolutionary algorithm has improved the initial design to better meet the turbine specifications. They also reported that the indirect encoding of DMFFD has provided three possible mappings between the object-point genotypes and the final design: many to one, one to one and illegal design. In order to apply DMFFD for industry scale evolutionary design tasks successfully, the authors note, these mapping scenarios need to be analyzed and understood.

The indirect encoding method mentioned in Chapter 4 is a version of embryogeny representations, which is receiving a lot of attention in the EC community recently. Briefly, embryogeny refers to the conversion of the genotypes through a set of instructions to develop or grow the phenotypes. The instructions can be intrinsic to the representation, where they are defined within the genotype, or extrinsic, where an external algorithm is used to develop the phenotype according to the values defined by the genotype. DMFFD is an example of embryogeny representations with an external growth algorithm.

In Chapter 5, Steven Manos and Peter Bentley also used an embryogeny representation with an external (more complicated) growth algorithm in their genetic algorithm to design micro-structured polymer optical fibres (MPOF). Although this is not the first application of evolutionary algorithms using an embryogeny representation, their solutions are innovative and patentable. This work received the Gold award at the 2007 Genetic and Evolutionary Computation Conference Human-Competitive Results competition. This applied research validates the practical value of the embryogeny representation in EC. We can anticipate that more EC applications will adopt this representation in the future.

Micro-structured optical fibres (MOF) are fibres which use air channels that run the length of the fibre to guide light. It is a newly developed technology that is being utilized in various optical communications, such as cable TV and operating theaters. The major task of MOF design is to position the air channels (air holes and their structures) such that they produce the required optical effects for a particular application, such as dispersion flattened fibres. This design task becomes very challenging when the material used is polymer, which allows any arbitrary arrangement of air holes.

To design MPOF, the authors devised a variable length genotype which encodes the positions and sizes of a variable number of air holes and the rotational symmetry of the structures. The conversion of the genotype into a MPOF is by a 4-step decoding algorithm, which not only transforms but also validates the design to satisfy manufacturing specifications and constraints, e.g. each hole must be surrounded by a minimum wall thickness for structural stability. They applied their genetic algorithm with the devised representation to design three different types of MPOF. The evolved designs are interesting and novel. Those that are of particular interest are currently being patented.

While the above work demonstrated the innovative aspect of EC, which can create human competitive designs, there are other areas of design in which human creativity is used to assist in the innovation. One example is consumer product design, such as the design of wallpapers and fabric tiles, where individual tastes play an important role in product acceptance. To make such collaboration possible, Chapter 6 discusses the development of an interactive evolutionary computation tool that can assist professional graphic designers to explore design patterns that go beyond their own imagination.

An interactive EC system opens the evolutionary algorithm's selection-reproduction-evaluation cycle and lets users guide the population toward a particular kind of solution. While the interface between users and an EC tool can have various degrees of flexibility, the one that motivates Carol Anderson, Daphna Buchsbaum, Jeff Potter and Eric Bonabeau is practicality: what feature an EC design tool should have so that graphic designers are more likely to use it in a regular basis to perform their jobs. They have considered many user-friendly features, such as allowing users to seed an initial design population and to freeze parts of the design without being modified by the reproduction operation. They are continuing to enhance the tool so that it can receive a wider acceptance by professional graphic designers.

So far, all the discussed works are in the arena of evolutionary design. This is not surprising since design is one of the most successful applications of EC. However, there are other application areas where EC solutions have been successfully deployed. The next four chapters discuss EC applications in business operations to reduce costs and increase revenues.

In Chapter 7, Cem Baydar applied EC-related technologies to grocery store retail operations, particularly in product pricing. In retail management, a product pricing strategy can impact store profits, sales volume and customer loyalty. A store manager can alter product prices by issuing coupons to achieve target objectives. One simple coupon-issuing strategy is blanket couponing, which offers the same discounts to every customer. In this chapter, Cem proposed an individual-pricing strategy where different discounts are offered to different customers, according to their buying behavior. His simulation results indicated that individual-pricing led to better store performance than the blanket couponing approach.

This result is not surprising, as more focused marketing normally produces results that meet a store's target goals better. However, the task of mapping this untraceable problem (the size of the search space is all possible combinations of coupon values and production items for each customer) into a workable framework using agent-based simulation and population-based simulated annealing is highly non-trivial. The problem-framing skill demonstrated in this project is important for practitioners who deliver real-world EC solutions.

Chapter 8 describes the implementation of a decision support tool for a routine railway operation: railway track intervention planning. Currently, the maintenance and renewal of railway track is mostly carried out by track maintenance engineers assisted by rule-based expert systems. Since the knowledge base of the expert systems is created by human engineers, who are not able to consider all combinations of possible track deterioration mechanisms, the resulting intervention plans are not very reliable and can lead to unnecessary high cost and low levels of safety.

To overcome such shortcomings, Derek Bartram, Michael Burrow and Xin Yao have proposed a data-driven approach to develop the decision support system. Based on historical data of railway track installation, deterioration, maintenance and renewal, they applied machine learning techniques to identify the patterns of failure types, to model the deterioration for each failure type and to determine the most appropriate maintenance for each failure type. The machine learning techniques they used include clustering, genetic algorithms and heuristic learning. Integrating multiple methods to address different needs of a system is becoming the standard approach for solve increasingly complex business problems. Practitioners need to develop the skill of picking the right technology in order to best solve the problem at hand.

Similar to the railway industry, there are areas in petroleum industry operations which rely on expert knowledge. One example is reservoir stratigraphic interpretation based on well log data, which is normally carried out by geologists or geophysicists who are familiar with the field. In Chapter 9, Tina Yu proposed a methodology using computer systems to automate this process.

Mimicking the ways geologists interpret well logs data, the proposed method has two steps: well logs blocking (approximation) and combining multiple logs' information at the same depth level to interpret reservoir properties. In terms of implementation, well logs blocking is carried out by a segmentation algorithm. Each block is assigned a fuzzy symbol to represent its approximate value. To interpret reservoir properties, a fuzzy rule set is generated to examine multiple well logs symbols at the same depth level and determine the property value. The prototype system also integrates various techniques, including a segmentation algorithm, fuzzy logic and a co-evolutionary system, to develop different parts of the interpretation system. Although the initial results based on well log data collected from two West Africa fields are encouraging, the author noted that there is weakness in the co-evolutionary system. More investigation is needed in order to produce quality deployable computer-based stratigraphic interpreters.

Resource scheduling is the task of allocating limited resources to requests during some period of time. In Chapter 10, Darrell Whitley, Andrew Sutton, Adele Howe and Laura Barbulescu discussed permutation based representations and their implementation with a genetic algorithm to solve three different resource scheduling problems. One interesting observation is that the permutations (genotypes) are indirect representations of the schedules (phenotypes). Different mapping algorithms were developed to transform permutations to schedules for different scheduling problems. This representation style is similar to the embryogeny technique presented in Chapter 5, although the work reviewed in this chapter was conducted at an earlier date when the term was not familiar to most EC technologists. Additionally, the authors have discussed various ways that the mapping algorithms, or schedule builder algorithms, can impact the success of their genetic algorithm to find good schedules. Much of the current research in embryogeny techniques are investigating similar issues.

2.   TECHNOLOGY TRANSFER

Technology transfer goes beyond applied research and focuses on the soft side of solution deployment, such as social and political issues. In Chapter 11, Arthur Kordon discusses transferring EC technology in corporate environments, such as Dow Chemical.

The chapter starts with a list of values EC contributes to the chemical industry. However, promoting the technology not only requires demonstrating value creation and improved performance but also resolving other non-technical issues. Currently, Dow has one of the most successful teams in deploying EC solutions in the corporate world. Arthur Kordon shared their experiences in how they handle the organization and political challenges to maintain their status in the company. Examples are linking EC to corporate initiatives and addressing skepticism and resistance toward EC technology. These insights are valuable to practitioners who are interested in establishing a sustainable EC team in corporate environments.

Another technology transfer model is the collaboration between academia and industry. With their extensive experience in this type of collaboration, Rajkumar Roy and Jorn Mehnen discussed this model in Chapter 12.

Collaboration between academia and industry is not a new concept. When carried out properly, it produces many rewards: universities receive funding to educate high-caliber students and to develop new technology, while industry receives ready-to-use state-of-art technology that increases company revenues, in addition to receiving highly qualified employees. However, because the two ends of the technology transfer belong to different organizations, trust between the two entities is not always easy to maintain, and lack of trust can cause project failures. In this chapter, the authors listed activities to create and maintain that trust throughout the project cycle. These include documenting expectations, conducting frequent review meetings and providing small deliverables regularly. Finally, the sensitive issue of intellectual property has to be addressed so that both parties can fully commit to the collaborative efforts.

When talking about technology transfer, the most direct way is placing trained EC technologists in jobs that demand EC technology. But how do these two find each other? Chapter 13 is a survey of practitioners of EC to identify potential job sectors and help EC graduates to do job searching.

The survey was conducted from March 2005 to February 2006 by Gregory Hornby and Tina Yu. The major findings from the survey data are: there has been an exponential growth in both EC graduates and practitioners; the main source for finding a job has been networking; while most respondents to the survey are in Europe, the most growth of EC in industry has been in North America; the main application areas of EC techniques are multi-objective optimization, classification, data mining and numerical optimization; and the biggest obstacle for the acceptance of EC techniques in industry is that it is poorly understood. This information, although it cannot be generalized to the entire population of EC practitioners, does provide some direction on where and how to search for an EC job.

One of the most common avenues to transfer emerging technologies, such as EC, to industry and government applications is external consultancy. Since EC was in its infancy in the early 1980s, Lawrence Davis has been working as an external consultant involving many EC projects. With his accumulated 25 years of consultancy experience, he shared twelve lessons learned to improve a project's chances of ultimate success in Chapter 14.

Among the 12 lessons, some are similar to those mentioned in the two previous chapters discussing technology transfer in corporate and in academia-industry collaboration. However, there are others which are unique in consultancy situations. They are mostly related to project perception management, which is a skill most technologists don't learn in academia. Yet, they are critical for a project to obtain funding, gain acceptance by the users and bring to final success. For example, have a project champion and don't speak technically make the project more accessible to non-technical managers. Understand the work process and the system's effect on it prepares the end-users to become familiar with the system and gains their support of final deployment. These non-technical skills are important for practitioners who are interested in consultancy for industrial applications.

With all the effort put forward on applied research and technical transfer, the statistics of successfully deployed EC systems are still very low: 10% according to Lawrence Davis. However, this does not mean that the time and energy spent on the rest of 90% projects were wasted. On the contrary, much has been gained throughout the projects. Davis calls it the Tao of optimization: it can be very valuable to improve company operations and increase revenues without deploying any computer systems. One example is by changing the workflow of a particular operation to reduce expenses.

For our final chapter we have included a success story of deployed EC systems, which have saved Air Liquide millions in operational expenses.

Air Liquide is a multi-national company providing industrial and medical gases and related services. One type of product they supply is liquid gases (oxygen, nitrogen and argon), which are delivered by truck. To coordinate the production and distribution of liquid gases, a system was developed with two parts: a genetic algorithm to schedule production and an ant colony algorithm to schedule distribution. The two systems co-evolve to obtain an integrated schedule that works together to generate the best result. This co-operative co-evolution approach is very effective in solving this complex supply-chain optimization problem. Charles Harper, the Director of National Supply and Pipeline Operations at Air Liquide Large Industries U.S.LP, reported that one of their plants has reported a saving of more than 1.5 million dollars per quarter since Air Liquide began using the system.

The second deployed system optimizes Air Liquide's gas pipeline operation. It uses a combination of techniques, including genetic algorithms, a deterministic heuristic algorithm and brute force search, to address different optimization issues in each sub-system. Using the right technology, as advocated by Lawrence Davis in the previous chapter, enables the team to create a system that generates outstanding performance.

3.   CONCLUSION

When we see EC systems working in the field to help people do their jobs better, it brings us a different kind of satisfaction from that gained by doing theorem proving. Regardless of whether you have created EC applications, we hope after reading this book, you will find applied research and technology transfer exciting and we hope that some of the insights in this book will help you to create and field high-impact EC systems.

Chapter 2

DESIGN FOR PRODUCT EMBEDDED DISASSEMBLY

Shingo Takeuchi[1] and Kazuhiro Saitou[1]

[1]*Department of Mechanical Engineering, University of Michigan, Ann Arbor, Michigan 48109-2125, USA*

Abstract   This chapter discusses an application of multi-objective genetic algorithm for designing products with a built-in disassembly means that can be triggered by the removal of one or a few fasteners at the end of the product lives. Given component geometries, the method simultaneously determines the spatial configuration of components, locators and fasteners, and the end-of-life (EOL) treatments of components and subassemblies, such that the product can be disassembled for the maxim profit and minimum environmental impact through recycling and reuse via domino-like "self-disassembly" process. A multi-objective genetic algorithm is utilized to search for Pareto optimal designs in terms of 1) satisfaction of the distance specification among components, 2) efficient use of locators on components, 3) profit of EOL scenario, and 4) environmental impact of EOL scenario. The method is applied to a simplified model of the Power Mac G4 cube® for demonstration.

Keywords: Design for disassembly, environmentally-conscious design, design optimization, multi-objective genetic algorithm

1.   INTRODUCTION

Increased regulatory pressures (e.g., EU's WEEE directive) and voluntary initiatives have placed manufacturers more responsible for end-of-life (EOL) treatments such as material recycling and component reuse. Since both recycling and reuse typically require disassembly, design for disassembly (DFD) has become a key design issue in almost any mass-produced product. DFD is particularly critical in consumer electronic products due to the large amount of production and short cycle time for technological obsolescence. Also, components

S. Takeuchi and K. Saitou: *Design for Product Embedded Disassembly*, Studies in Computational Intelligence (SCI) **88**, 9–39 (2008)
www.springerlink.com

in these products are typically required to fit into a tight enclosing space, which makes disassembly even more challenging.

Economic feasibility of an end-of-life (EOL) scenario of a product is determined by the interaction among disassembly cost, revenue from the EOL treatments of the disassembled components, and the regulatory requirements on products, components and materials. While meeting regulatory requirements is obligatory regardless of economic feasibility, EOL decision making is often governed by economical considerations (Chen et al., 1993). Even if a component has high recycling/reuse value or high environmental impact, for instance, it may not be economically justifiable to retrieve it if doing so requires excessive disassembly cost. Since the cost of manual disassembly depends largely on the number of fasteners to be removed and of components to be reached, grabbed, and handled during disassembly, it is highly desirable to locate such high-valued or high-impact components within a product enclosure, such that they can be retrieved by removing less fasteners and components.

The above thoughts motivated us to develop a concept of product-embedded disassembly, where the relative motions of components are constrained by locator features (such as catches and lugs) integral to components, in such a way that the optimal disassembly sequence is realized via a domino-like "self-disassembly" process triggered by the removal of one or a few fasteners.

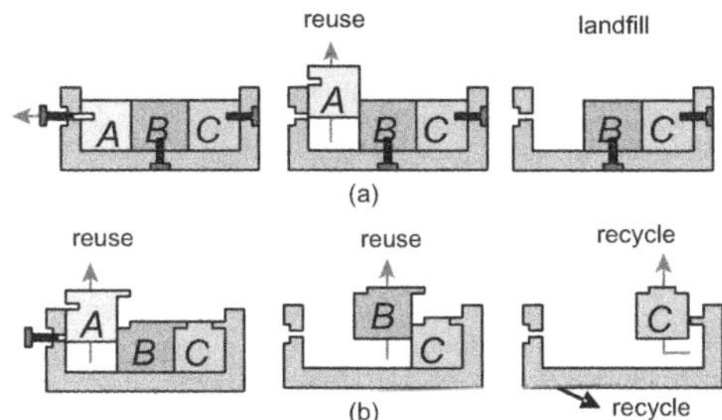

Figure 2-1.   (a) Conventional assembly (b) assembly with embedded disassembly.

Fig. 2-1 illustrates the concept of product-embedded disassembly as compare to the conventional disassembly. In the conventional assembly (Fig. 2-1a), components A, B and C are fixed with three fasteners. With high labor cost for removing fasteners (as often the case in developed countries), only

A may be economically disassembled and reused, with the remainder sent to a landfill. This end-of-life (EOL) scenario (*i.e.*, disassemble and reuse A, and landfill the remainder) is obviously not ideal from either economical or environmental viewpoints. In the assembly with embedded disassembly (Fig. 2-1b), on the other hand, the motions of B and C are constrained by the locators on components. As such, the removal of the fastener to A (called a trigger fastener) activates the domino-like self-disassembly pathway A $\rightarrow$ B $\rightarrow$ C. Since no additional fasteners need to be removed, B and C can also be disassembled, allowing the recycle/reuse of all components and the case. This EOL scenario (*i.e.*, disassemble all components, reuse A and B, and recycle C and the case) is economically and environmentally far better than the one for the conventional assembly.

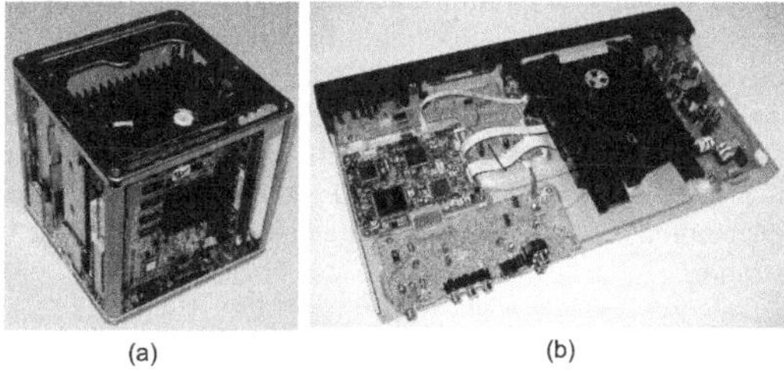

Figure 2-2. Example products suited for product-embedded disassembly, (a) desktop computer (b) DVD player.

The concept of product-embedded disassembly can be applied to a wide variety of products, since it requires no special tools, materials, or actuators to implement. It is particularly well suited for electrical products assembled of functionally modularized components, whose spatial configurations within the enclosure have some flexibility. Fig. 2-2 shows examples of such products. A desktop computer in Fig. 2-2a is assembled of functionally distinct components such as a motherboard, a hard drive, and a power unit, arranged to fit within a tight enclosure. The components are, however, not completely packed due to the need of the air passage for cooling and the accessibility for upgrade and repair. Thanks to this extra space and electrical connections among components, the spatial configurations of the components have a certain degree of flexibility. A DVD player in Fig. 2-2b shows even roomier component arrangements, due to the consumers' tendency to prefer large sizes in home theater appliances. Since designing products with a single "disassembly button" may cause safety concerns, the method will, in practice, be best utilized as an inspiration to the designer during the early stage configuration design and critical components can be independently fastened with a secure, conventional means.

The concept, however, may be unsuitable to the products that allow very little freedom in component arrangements. Examples include mobile IT products such as cell phones, laptop computers, and MP3 players, due to their extremely tight packaging requirements and mostly layer-by-layer assembly.

This chapter discusses an application of a multi-objective genetic algorithm for designing products that optimally embody the above concept of product-embedded disassembly (Takeuchi and Saitou, 2005; Takeuchi and Saitou, 2006). Given component geometries, the method simultaneously determines the spatial configuration of components, locators and fasteners, and the end-of-life (EOL) treatments of components and subassemblies, such that the product can be disassembled for the maximum profit and minimum environmental impact through recycling and reuse via a domino-like "self-disassembly" process. A multi-objective genetic algorithm (Fonseca and Fleming, 1993; Deb et al., 2002) is utilized to search for Pareto optimal designs in terms of 1) satisfaction of the distance specification among components, 2) efficient use of locators on components, 3) profit of EOL scenario, and 4) environmental impact of EOL scenario. The method is applied to a simplified model of Power Mac G4 cube® for demonstration.

2.   RELATED WORK

2.1   Design for Disassembly

Design for disassembly (DFD) is a class of design methods and guidelines to enhance the ease of disassembly for product maintenance and/or EOL treatments (Boothroyd and Alting, 1992). Kroll et al. (1996) utilized disassembly evaluation charts to facilitate the improvements of product design. Das et al. (2000) introduced the Disassembly Effort Index (DEI) score to evaluate the ease of disassembly. Reap and Bras (2002) reported DFD guidelines for robotic semi-destructive disassembly, where detachable or breakable snap fits are preferred to screws due to their ease of disengagement. O'Shea et al. (1999) focused on tool selection during disassembly where the optimal tool selection path, in terms of the ease of disassembly, is produced via dynamic programming. Recently, Desai and Mital (2003) developed a scoring system, where factors associated with disassembly time such as disassembly force, tool requirements, and accessibility of fasteners are considered. Sodhi et al. (2004) focused on the impact of unfastening actions on disassembly cost and constructed U-effort model that helps designers to select fasteners for easy disassembly. Matsui et al. (1999) proposed the concept of Product Embedded Disassembly Process, where a means of part separation that can be activated upon disassembly is embedded within a product. As an example, they developed cathode-ray tube (CRT) with a Nichrome wire embedded along the desired separation line, which can induce thermal stress to crack the glass of the CRT tube upon the application of current.

While these works suggest locally redesigning an existing assembly for improving the ease of its disassembly, they do not address the simultaneous decisions of the spatial configuration of components and joints for improving the entire disassembly processes.

2.2 Disassembly Sequence Planning

Disassembly Sequence Planning (DSP) aims at generating feasible disassembly sequences for a given assembly, where the feasibility of a disassembly sequence is checked by the existence of collision-free motions to disassemble each component or subassembly in the sequence. Since the disassembly sequence generation problem is NP-complete, the past research has focused on efficient heuristic algorithms to approximately solve the problem. Based on a number of important research results on assembly sequence planning (Homem dé Mello and Sanderson, 1990; De Fazio and Whitney, 1987; Lee and Shin, 1990; Homem dé Mello and Sanderson,. 1991; Baldwin, et al., 1992), several automated disassembly sequence generation approaches for 2/2.5D components have been developed (Woo and Dutta, 1991; Dutta and Woo, 1995; Chen et al., 1997; Srinivasan and Gadh, 2000; Kaufman et al., 1996). More recent work is geared towards DSP with special attention to reuse, recycling, remanufacturing, and maintenance. Lambert (1999) built a linear programming model to obtain the optimal EOL disassembly. Li et al. (2002) used Genetic Algorithm (GA) combined with Tabu search (Glover, 1974; Glover 1986) to find the optimal disassembly sequence for maintenance.

This previous work, however, only addresses the generation and optimization of disassembly sequences for an assembly with a pre-specified spatial configuration of components. Since the feasibility of disassembly sequences largely depends on the spatial configuration of components, this would seriously limit the opportunities for optimizing an entire assembly. In addition, these works do not address the design of joint configurations, which also have a profound impact on the feasibility and quality of a disassembly sequence.

2.3   Configuration Design Problem

While rarely discussed in the context of disassembly, the design of the spatial configuration of given shapes have been an active research area by itself. Among the most popular flavors is the bin packing problem (BPP), where the total volume (or area for 2D problems) a configuration occupies is to be minimized. Since this problem is also NP-complete, heuristic methods are commonly used. Fujita et al. (1996) proposed hybrid approaches for a 2D plant layout problem, where the topological neighboring relationships of a layout are determined by Simulated Annealing (SA), whereas the generalized reduced gradient (GRG) method determines the geometry. Kolli et al. (1996) used SA for packing 3D components with arbitrary geometry. GA is also widely used for the configuration design problem. Corcoran and Wainwright (1992) solved a 3D packing problem with GA using multiple crossover methods. Jain and Gea (1998) adopted discrete representation as an object expression and proposed a geometry-based crossover operation for a 2D packing problem. Grignon and Fadel (1999) proposed a configuration design optimization method by using multi-objective GA, where static and dynamic balances and maintainability are considered in addition to configuration volume.

These works, however, do not address the integration with DSP.

2.4   Life Cycle Assessment

Life Cycle Assessment (LCA) has been widely used as a tool to estimate the environmental impact of an EOL scenario of various products (Caudill et al., 2002; Rose and Stevels, 2001) including computers (Williams and Sasaki, 2003; Aanstoos et al., 1998; Kuehr and Williams, 2003). Since the optimal EOL scenario should be economically feasible as well as environmentally sound, LCA is often integrated with cost analysis. Goggin and Browne (2000) constructed a model for determining the recovery of a product, components and materials, where EOL scenarios are evaluated from economical and environmental perspectives. Kuo and Hsin-Hung (2005) integrated LCA into Quality Function Development (QFD) to achieve the best balance between customer satisfaction and environmental impact. In our previous work (Hula et al., 2003), we compared the optimal EOL scenarios of a coffee maker in Aachen, Germany and in Ann Arbor, MI, and concluded the optimal EOL scenario varied greatly depending on the local recycling/reuse infrastructures and regulatory requirements.

This work, however, merely addressed the evaluation and optimization of the environmental impact of a given product, and did not address the design of component, locator, and fastener configuration as addressed in this paper.

3. METHOD

The method can be summarized as the following optimization problem:

- **Given**: geometries, weights, materials, and recycle and reuse values of each component, contact and distance specifications among components, locator library, and possible EOL treatments and associated scenarios.
- **Find**: spatial configuration of components and locators, EOL treatments of disassembled components and subassemblies.
- **Subject to**: no overlap among components, no unfixed components prior to disassembly, satisfaction of contact specifications, assemble-ability of components.
- **Minimizing**: violation of distance specification, redundant use of locators, and environmental impact of EOL scenario.
- **Maximizing**: profit of EOL scenario.

Since the optimization problem has four objectives, a multi-objective genetic algorithm (MOGA) (Fonseca and Fleming, 1993; Deb et al., 2002) is utilized to obtain Pareto optimal solutions.

3.1 Inputs

There are four (4) categories of inputs for the problem as listed below:

- **Component information**: This includes the geometries, weights, materials and reuse values of components. Due to the efficiency in checking contacts and the simplicity in modifying geometries (Beasley and Martin, 1993; Minami et al., 1995; Sung et al., 2001), the component geometries are represented by voxels. CAD inputs are first voxelized using ACIS® solid modeling kernel.
- **Contact and distance specifications**: The adjacencies and distances among components are often constrained by their functional relationships. For example, a heat sink and CPU in a computer should be in contact, and a cooling fan and CPU should be nearby. The contact specification specifies the required adjacencies among the component, such as CPU and a heat sink in a computer. Since the distances between some pairs of components are more important than the others, the distance specification is defined as a set of the weights of importance for the distances between pairs of components (measured between two designated voxels) that need to be minimized. If the weight between two components is not defined, their distance is considered unimportant and can be arbitrarily chosen. Fig. 2-3 shows an example.

- **Locator library**: Since the types of feasible locators depend on manufacturing and assembly processes, they are pre-specified by a designer as a locator library. It is a set of locators for a specific application domain, which can be potentially added on each component to constrain its motion. Fig. 2-4 shows schematics of locators commonly found on sheet metal or injection-molded components in computer assemblies (Bonenberger, 2000), which are also used in the following case study. Note that screws are regarded as a special type of locators, and a slot can only be used with two circuit boards. Locator constraints (LC) shown in the third column of Fig. 2-4 illustrates the set of directions locators are constrained to when they are oriented as shown in the second column, formally represented as a subset of $\{-x, +x, -y, +y, -z, +z\}$.

- **Possible EOL treatments and scenarios**: An EOL scenario is a sequence of events, such as disassembly, cleaning, and refurbishing, before a component receives an EOL treatment such as recycle and reuse. The EOL treatments available to each component and the associated scenarios leading to each treatment must be given as input. Fig. 2-5 shows an example of EOL treatments (reuse, recycle, or landfill) and the associated EOL scenarios represented as a flow chart.

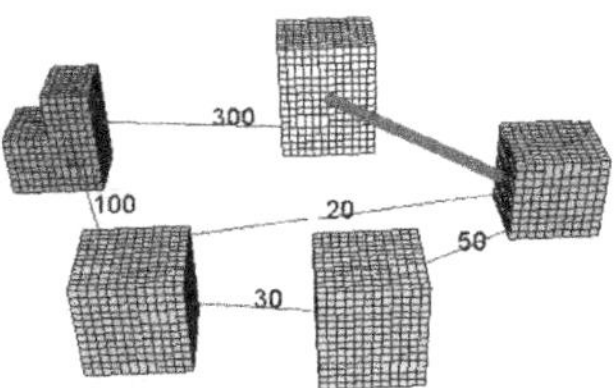

Figure 2-3.   Example of contact specification (thick line) and distance specification (thin lines). Labels on thin lines indicate relative importance of minimizing distances.

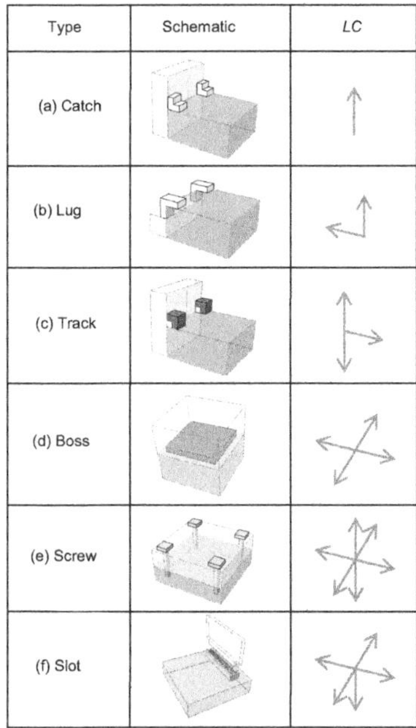

Figure 2-4. Graphical representation of typical locators for sheet metal or injection-molded components (Bonenberger, 2000).

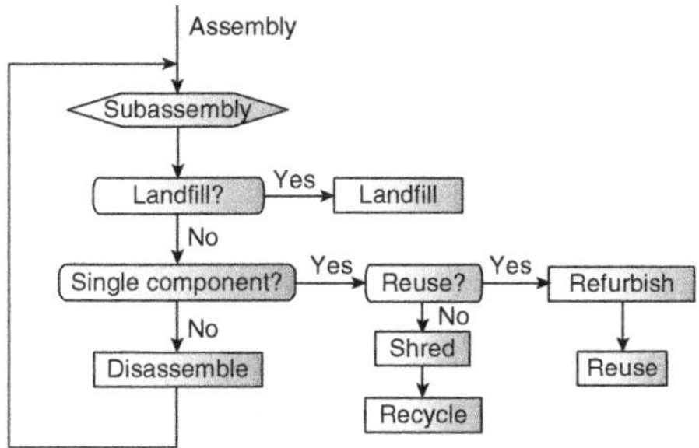

Figure 2-5. Flow chart of example EOL scenario.

3.2   Design Variables

There are three (3) design variables for the problem. The first design variable, *configuration vector*, represents the spatial configuration and dimensional change of each component:

$$\boldsymbol{x} = (x_0, x_1, \ldots, x_{n-1}) \tag{1}$$

$$x_i = (\boldsymbol{t}_i, \boldsymbol{r}_i, \boldsymbol{d}_i,); \; i = 0, 1, \ldots n - 1 \tag{2}$$

$$\boldsymbol{t}_i \in \{0, \pm c, \pm 2c, \pm 3c, \ldots\}^3 \tag{3}$$

$$\boldsymbol{r}_i \in \{-90°, 0°, 90°, 180°\}^3 \tag{4}$$

$$\boldsymbol{d}_i \in \{0, \pm c, \pm 2c, \pm 3c, \ldots\}^f \tag{5}$$

where n is the number of components in the assembly, $\boldsymbol{t}_i$ and $\boldsymbol{r}_i$ are the vectors of the translational and rotational motions of component i with respect to the global reference frame, and $\boldsymbol{d}_i$ is a vector of the offset values of the f faces of component i in their normal directions, and c is the length of the sides of a voxel. Note that $\boldsymbol{d}_i$ is considered only for the components whose dimensions can be adjusted to allow the addition of certain locator features. For example, the components designed and manufactured in-house can have some flexibility in their dimensions, whereas off-the-shelf components cannot.

The second design variable, *locator vector*, represents the spatial configuration of the locators on each component:

$$\boldsymbol{y} = (\boldsymbol{y}_0, \boldsymbol{y}_1, \ldots, \boldsymbol{y}_{m-1}) \tag{6}$$

$$\boldsymbol{y}_i = (CD_i, p_i); \; i = 0, \ldots, m - 1 \tag{7}$$

where $m = n(n - 1)/2$ is the number of pairs of components in the assembly, and $CD_i \subseteq \{-x, +x, -y, +y, -z, +z\}$ is a set of directions in which the motion of component c_0 in the i-th pair (c_0, c_1) is to be constrained, and p_i is a sequence of locators indicating their priority during the construction of the locator configuration.

The choice of locator for the i-th component pair is indirectly represented by CD_i and p_i, since the direct representation of the locator id in the library would result in a large number of infeasible choices. The construction of locator configurations from a given y_i is not trivial since 1) multiple locator types can constrain the motion of c_0 as specified by CD_i, and 2) among such locator types, geometrically feasible locators depend on the relative locations of components c_0 and c_1. Fig. 2-6 shows an example. In order to constrain the motion of c_0 in $+z$ direction, a catch can be added to c_1 if c_0 is "below" c_1 as shown in Fig. 2-6a. However, a catch cannot be used if c_0 is "above" c_1 as shown in Figs. 2-6b and 6c, in which case boss (Fig. 2-6b) or track (Fig. 2-6c) needs to be used. Thus, the locator configuration of a component is dynamically constructed by testing locator types in the sequence of p_i.

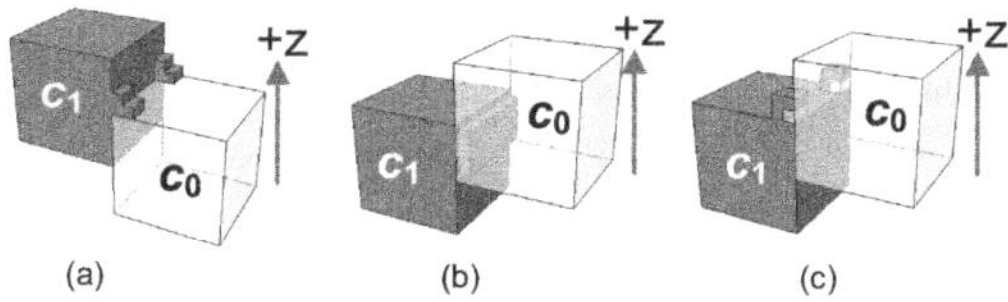

Figure 2-6. Construction of locator configuration.

Given $y_i = (CD_i,\ p_i)$, the locator configuration of the i-th pair of components c_0 and c_1 is constructed by testing locator types, in sequence p_i, for constraining each direction in CD_i as follows:

1. For each $d \in CD_i$, remove d from CD_i if the motion of c_0 in $d \in CD$ is constrained by other components or locators. This step is necessary to reduce the redundant use of locator features.
2. Remove locator type t at the beginning of p_i. If p_i is empty, return FALSE.
3. Select direction $d \in CD$.
4. Find an orientation of o of locator type t whose locator constraint LC (after re-orientation) contains d. If several orientations are found, select an orientation with maximum $|LC \cap CD_i|$. If none is found, go to step 2.
5. Add t to c_0 or c_1 in o.
6. $CD_i \leftarrow CD_i \backslash LC$. If $CD = \varnothing$, return TRUE. Otherwise, go to step 3.

The above procedure returns TRUE if a locator configuration constraining all directions in CD_i is found by using the locator types in p_i, and FALSE otherwise. During optimization, the value of $\boldsymbol{y}_i$ returning FALSE is considered as infeasible.

Fig. 2-7 shows an example construction of locator configuration of components c_0 and c_1 according to the above procedure with $CD = \{+z\}$ and $p = $ <Catch, Screw, Lug, Track, Boss>:

- Step 1: Since component c_1 does not constrain the motion of c_0 in $+z$ (Fig. 2-7a), $+z$ remains in CD

- Step 2: Remove Catch from p. Since $p = $ <Screw, Lug, Track, Boss> is non-empty, proceed.

- Step 3: Select $+z$ from CD.

- Step 4: Systematically examine the possible orientations of Catch on c_0 and c_1 to find the orientations that constraint $+z$ (o_0 through o_7 in Fig. 2-7b and 7c). Note, however, that the orientations other than o_0 and o_5 in Fig. 2-7d are infeasible due to the lack of an adjacent component. Since both o_0 and o_5 has $|LC \cap CD_i| = |\{+z\} \cap \{+z\}| = |\{+z\}| = 1$, o_0 is chosen.

- Step 5: Catch in orientation o_0 is added to c_1 (Fig. 2-7e).

- Step 6: Since $CD_i \backslash LC = \{+z\} \backslash \{+z\} = \emptyset$, $CD_i = \emptyset$. Return TRUE.

Fig. 2-8 illustrates how two different values of priority sequence p with the same CD can result in the different locator configurations. For the two components in Fig. 2-8a with $CD = \{-x, +x, +z\}$, sequence $p = $ <Track, Boss, Screw, Catch, Lug> results in the locators in Fig. 2-8b, whereas sequence $p = $ <Catch, Lug, Screw, Track, Boss> results in the locators in Fig. 2-8c. In Fig. 2-8c, two locator types, Catch and Lugs are used since Catch (top priority) cannot be oriented to constrain c_0 in $+z$ direction while Lug (second priority) can.

While indirect, constraint direction CD and priority sequence p realizes a compact representation of a locator configuration of a pair of components. Compared to the direct representation in (Takeuchi and Saitou, 2005) that specifies the existence of a locator type in an orientation at a potential location on a component, it can generate far fewer infeasible locator configurations during the "generate and test" process of genetic algorithms. As a result, the computational efficiency is dramatically improved. Instead of treating the priority sequence as a design variable, one might imagine checking for locator types always in the (fixed) ascending sequence of their manufacturing costs is sufficient. However, such costs are difficult to determine *a priori*, since the actual geometry (and hence the cost) of a locator heavily dependents on the configuration of the surrounding components.

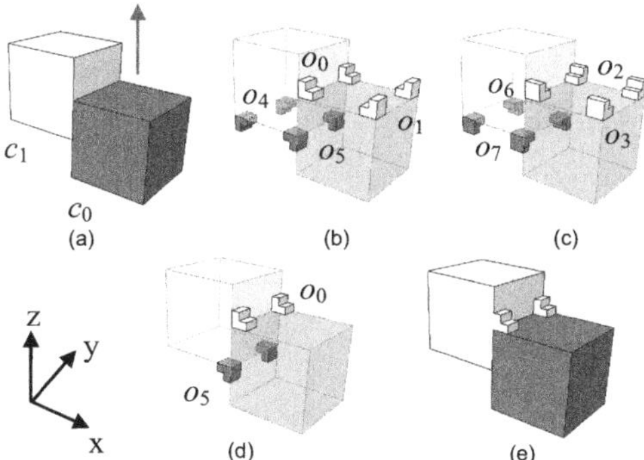

Figure 2-7. An example construction of locator configuration: (a) two components, (b) and (c) possible orientations of Catch, (d) two feasible orientations, and (e) final locator configuration.

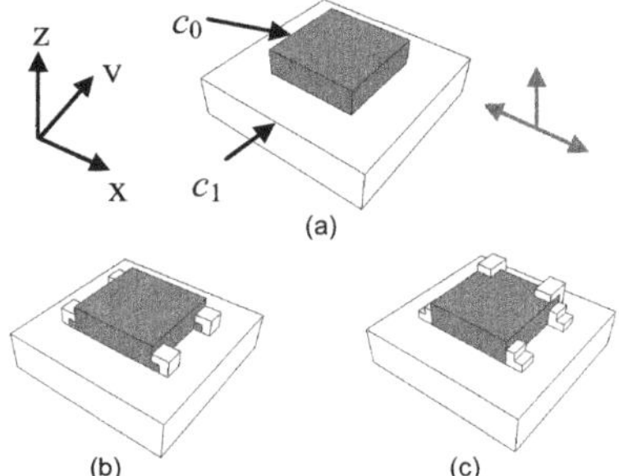

Figure 2-8. Influence of priority sequence p in locator configurations: (a) two components with $CD = \{-x, +x, +z\}$, (b) locators constructed with $p = <$Track, Boss, Screw, Catch, Lug$>$, and (c) locators constructed with $p = <$Catch, Lug, Screw, Track, Boss$>$.

The third design variable, *EOL vector*, represents the EOL treatments of components:

$$z = (z_0, z_1, \ldots, z_{n-1}); z_i \in E_i \qquad (8)$$

where E_i is a set of feasible EOL treatments of component i. In the following case study, $E_i = \{$recycle, reuse, landfill$\}$ for all components.

3.3   Constraints

There are four (4) constraints for the problem:

1. No overlap among components.
2. Satisfaction of contact specification.
3. No unfixed components prior to disassembly.
4. Assemble-ability of components.

Since the constraints are all geometric in nature, the voxel representation of component geometry facilitates their efficient evaluation. Constraints 1-3 are checked solely based on the information in x, since the locator configurations constructed from y generate no overlaps. For constraint 3, immobility of all possible subassemblies is examined. Constraint 4 is necessary to ensure all components, whether or not to be disassembled, can be assembled when the product is first put together. It requires the information from both x and y. Since checking this constraint requires simulation of assembly motions (assumed as the reverse of disassembly motions), it is done as a part of the evaluation of disassembly cost needed for one of the objective functions.

3.4   Objective Functions

There are four (4) objective functions for the problem. The first objective function (to be minimized) is for the satisfaction of the distance specification, given as:

$$f_1(\boldsymbol{x}, \boldsymbol{y}) = \sum_i w_i d_i \tag{9}$$

where w_i is the weight of the importance of distance d_i between two designated voxels.

The second objective function (to be minimized) is for the efficient use of locators, given as:

$$f_2(\boldsymbol{x}, \boldsymbol{y}) = \sum_i mc_i \tag{10}$$

where mc_i is the manufacturing difficulty of the i-th locator in the assembly, which represents the increased difficulty in manufacturing components due to the addition of the i-th locator.

The third objective function (to be maximized) is the profit of the EOL scenario of the assembly specified by x and y, given as:

$$f_3(x,y,z) = \sum_{i=0}^{n-1} p_i(z_i) - c^*(x,y,z) \tag{11}$$

In Eq. 11, $p_i(z_i)$ is the profit of the i-th component from EOL treatment z_i, calculated by the EOL model described in the next section. Also in Eq. 11, $c^*(x, y, z)$ is the minimum disassembly cost the assembly under the EOL scenario required by z:

$$c^*(x,y,z) = \min_{s \in S_{xyz}} c(s) \tag{12}$$

where S_{xyz} is the set of the *partial and total* disassembly sequences of the assembly specified by x and y, for retrieving the components with $z_i =$ reuse or recycle and the components with regulatory requirement, and $c(s)$ is the cost of disassembly sequence s. Since an assembly specified by x and y can be disassembled in multiple sequences, Eq. 12 computes the minimum cost over S_{xyz}, which contains all disassembly sequences feasible to x, y, and z, and their subsequences. Set S_{xyz} is represented as AND/OR graph (Homem dé Mello and Sanderson, 1990) computed based on the 2-disassemblability criterion (Woo and Dutta, 1991; Beasley and Martin, 1993) (i.e. the component can be removed by up to two successive motions) as follows:

1. Push the assembly to stack Q and the AND/OR graph.
2. Pop a subassembly sa from Q.
3. If sa does not contains component with $z_i =$ reuse or recycle and components with regulatory retrieval requirements, go to step 5.
4. For each subassembly $sb \subset sa$ that does not contain any fixed components, check the 2-disassemblability of sb from sa. If sb is 2-disassemblable, add sb and $sc = sa \backslash sb$ to the AND/OR graph. If sb and/or sc are composed of multiple components, push them to Q.
5. If $Q = \varnothing$, return. Otherwise go to step 2.

where the 2-disassembleability of two subassemblies sb from sa is checked as follows. For efficiency, only translational motions are considered:

1. For each mating surfaces between sb and $sc = sa \backslash sb$ (including the ones of the locators), obtain a set of constrained directions as a subset of six possible translational directions $D = \{-x, +x, -y, +y, -z, +z\}$.
2. Compute constrained directions CD_{bc} between sb and sb as a union of all constrained directions obtained in step 1.
3. If $D \backslash CD_{bc} = \emptyset$, return FALSE.
4. If there exist a direction in $D \backslash CD_{bc}$ along which sb can be moved infinitely without a collision, return *TRUE* (sb is 1-disassembleable).
5. Select a direction d in $D \backslash CD_{bc}$. If all have been selected, return infeasible. Otherwise, go to the next step.
6. Move sb by unit length along d. If sb collides with other components, go to step 5.
7. If sb is 1-disassembleable at the current location, return TRUE (sb is 2-disassembleable). Otherwise, go to step 6.

Assuming manual handling, insertion and fastening as timed in (Boothroyd et al., 1994), $c(s)$ is estimated based on the motions of the components and the numbers and accessibilities of the removed screws at each disassembly step. The cost of the i-th disassembly step is given by:

$$c_i = \omega_0 \cdot dc_0 + \omega_1 \cdot dc_1 + \omega_2 \cdot dc_2 \tag{13}$$

where dc_0 and dc_1 are the number of orientation changes and the sum of the moved distances, respectively, of the disassembled component at the i-th disassembly step, dc_2 is the sum of accessibilities a_s of the removed screws, and ω_j is the weights. The accessibility a_s of a removed screw is given as (Takeuchi and Saitou, 2005):

$$a_s = 1.0 + \omega_a/(aa + 0.01) \tag{14}$$

where ω_a is weight and aa is the area of the mounting face of the screw accessible from the outside of the product in its normal direction.

The forth objective function (to be minimized) is the environmental impact of the EOL scenario:

$$f_4(z) = \sum_i e_i(z_i) \tag{15}$$

where $e_i(z_i)$ is the environmental impact of i-th component according to the EOL scenario for treatment z_i. The value of $e_i(z_i)$ is estimated by the EOL model described in the next section.

3.5   EOL Model

The EOL model adopted in the following case study assumes the EOL scenarios in Fig. 2-5 for all components (reuse only for some components), and uses energy consumption as the indicator for environmental impact (Hula et al., 003). Accordingly, profit $p_i(z_i)$ in Eq. 11 is defined as:

$$p_i(z_i) = \begin{cases} r_i^{reuse} - c_i^{trans} - c_i^{refurb} & \text{if } z_i = \text{reuse} \\ r_i^{recycle} - c_i^{trans} - c_i^{shred} & \text{if } z_i = \text{recycle} \\ -c_i^{trans} - c_i^{landfill} & \text{if } z_i = \text{landfill} \end{cases} \qquad (16)$$

where r_i^{reuse} and $r_i^{recycle}$ are the revenues from reuse and recycle, respectively, and c_i^{trans}, c_i^{refurb}, c_i^{shred} and $c_i^{landfill}$ are the cost for transportation, refurbishment, shredding, and landfill, respectively. Similarly, energy consumption $e_i(z_i)$ in Eq. 15 is defined as:

$$e_i(z_i) = \begin{cases} e_i^{reuse} + e_i^{trans} + e_i^{refurb} & \text{if } z_i = \text{reuse} \\ e_i^{recycle} + e_i^{trans} + e_i^{shred} & \text{if } z_i = \text{recycle} \\ e_i^{landfill} + e_i^{trans} & \text{if } z_i = \text{landfill} \end{cases} \qquad (17)$$

where e_i^{reuse}, e_i^{trans}, $e_i^{recycle}$, e_i^{refurb}, e_i^{shred} and $e_i^{landfill}$ are the energy consumptions of reuse, transportation, recycle, refurbishment, shredding, and landfill, respectively.

Revenue from reuse r_i^{reuse} is the current market value of component i, if such markets exist. Energy consumption of reuse e_i^{reuse} is the negative of the energy recovered from reusing component i:

$$e_i^{reuse} = -\sum_j me_j^{intens} \cdot m_{ij} \qquad (18)$$

where me_j^{intens} is the energy intensity of material j and m_{ij} is the weight of material j in component i. Reuse, if available, is usually the best EOL treatment for a component because of its high revenue and high energy recovery. The availability of the reuse option for a component, however, is infrastructure dependent, and even if available, the revenue from reuse can greatly fluctuate in the market and hence difficult to estimate *a priori*.

Revenue from recycle $r_i^{recycle}$ and energy consumption of recycle $e_i^{recycle}$ are also calculated based on the material composition of a component:

$$r_i^{recycle} = \sum_j mr_j^{recycle} \cdot m_{ij} \tag{19}$$

$$e_i^{recycle} = -\sum_j me_j^{recover} \cdot m_{ij} \tag{20}$$

where $mr_j^{recycle}$ and $me_j^{recover}$ are the material value and recovered energy of material j, respectively.

Since little data is available for the refurbishment of components, the cost for refurbishment is simply assumed as:

$$c_i^{refurb} = 0.5 \cdot r_i^{reuse} \tag{21}$$

Based on the data on desktop computers (Aanstoos et al., 1998), energy consumption for refurbishment e_i^{refurb} is estimated as:

$$e_i^{refurb} = 1.106 \cdot m_i \tag{22}$$

where m_i is the weight of the i-th component.

Cost and energy consumption of transportation c_i^{trans} and e_i^{trans} are estimated as (Hula et al., 2003):

$$c_i^{trans} = \Delta c_i^{trans} \cdot D_i \cdot m_i \tag{22}$$

$$e_i^{trans} = \Delta e_i^{trans} \cdot D_i \cdot m_i \tag{23}$$

where $\Delta c_i^{trans} = 2.07\text{e} - 4\ [\$/\text{kg} \cdot \text{km}]$, $\Delta e_i^{trans} = 1.17\text{e} - 3\ [\text{MJ}/\text{kg} \cdot \text{km}]$, and D_i is the travel distance. Similarly, costs and energy consumptions for shredding and landfill c_i^{shred}, $c_i^{landfill}$, e_i^{shred} and $e_i^{landfill}$ are calculated as (Hula et al., 2003):

$$c_i^{shred} = \Delta c_i^{shred} \cdot m_i \tag{24}$$

$$e_i^{shred} = \Delta e_i^{shred} \cdot m_i \tag{25}$$

$$c_i^{landfill} = \Delta c_i^{landfill} \cdot m_i \tag{26}$$

$$e_i^{landfill} = \Delta e_i^{landfill} \cdot m_i \tag{27}$$

where $\Delta c_i^{shred} = 0.12\,[\$/\text{kg}\cdot\text{km}]$, $\Delta e_i^{shred} = 1.0\,[\text{MJ}/\text{kg}\cdot\text{km}]$, $\Delta c_i^{landfill} = 0.02\,[\$/\text{kg}\cdot\text{km}]$ and $\Delta e_i^{landfill} = 20000\,[\text{MJ}/\text{kg}\cdot\text{km}]$.

3.6   Optimization Algorithm

Since the problem is essentially a "double loop" of two NP-complete problems (i.e. disassembly sequence planning within a 3D layout problem), it should be solved by a heuristic algorithm. Since design variables x, y, z are discrete (x is a discrete variable since geometry is represented as voxels) and there are four objectives, a multi-objective genetic algorithm (Fonseca and Flemming, 1993; Deb et al., 2002) is utilized to obtain Pareto optimal design alternatives. A multi-objective genetic algorithm is an extension of the conventional (single-objective) genetic algorithms that do not require multiple objectives to be aggregated to one value, for example, as a weighted sum. Instead of static aggregates such as a weighted sum, it dynamically determines an aggregate of multiple objective values of a solution based on its relative quality in the current population, typically as the degree to which the solution dominates others in the current population.

A chromosome, a representation of design variables in genetic algorithms, is a simple list of the 3 design variables:

$$c = (x, y, z) \tag{28}$$

Since the information in x, y, and z are linked to the geometry of a candidate design, the conventional one point or multiple point crossover for linear chromosomes are ineffective in preserving high-quality building blocks. Accordingly, a geometry-based crossover operation based on (Jain and Gea, 1998) is adopted:

1. Randomly select a point in the bounding box of the assembly.
2. Cut two parent designs p_1 and p_2 with the three planes parallel to x, y, z axes, and passing through the point selected in step 1, into eight pieces each (Fig. 2-9a).
3. Assemble two child designs c_1 and c_2 by alternately swapping the pieces of p_1 and p_2 (Fig. 2-9b).
4. Repair c_1 and c_2 by moving each component C to the child containing the larger volume (of the sliced piece) of C. If c_1 and c_2 contain the same volume, C is placed in the same way as the parent with the higher rank.
5. Add locators to c_1 and c_2 by checking which parent each pair of component is inherited from. If a child contains both components of a pair, the corresponding locator is added to the child. Otherwise, a locator is randomly added to either child.

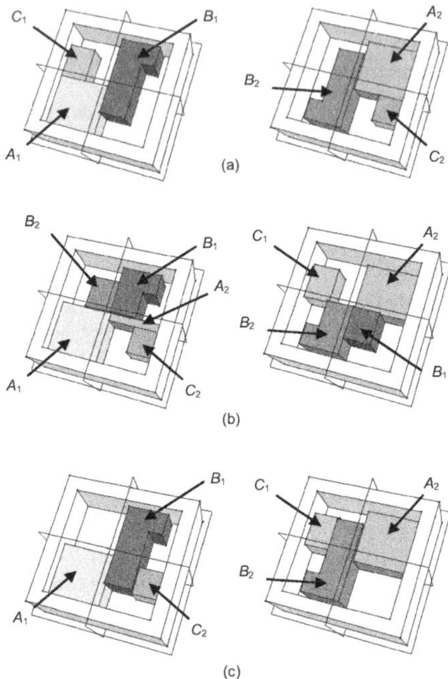

Figure 2-9.   Geometry-based crossover operator. (a) two parents p_1 (left) and p_2 (right), (b) two children c_1 (left) and c_2 (right) after crossover, and (c) two children c_1 (left) and c_2 (right) after repair.

4.   CASE STUDY

4.1   Problem

The method is applied to a model of Power Mac G4 Cube® manufactured by Apple Computer, Inc. (Fig. 2-10). Ten (10) major components are chosen based on the expected contribution to profit and environmental impact. Fig. 2-11a shows the ten components and their primary liaisons, and Fig. 2-11b shows the voxel representation of their simplified geometry and the contact (thick lines) and distance (thin lines with weights) specifications. The contacts between component B (heat sink) and C (CPU), and C (circuit board) and G (memory) are required due to their importance to the product function. Component A (case) is considered as fixed in the global reference frame. Component J (battery) needs to be retrieved due to regulatory requirements. The locator library in Fig. 2-4 is assumed for all components. The relative manufacturing difficulty of locators in the library is listed in Table 2-1.

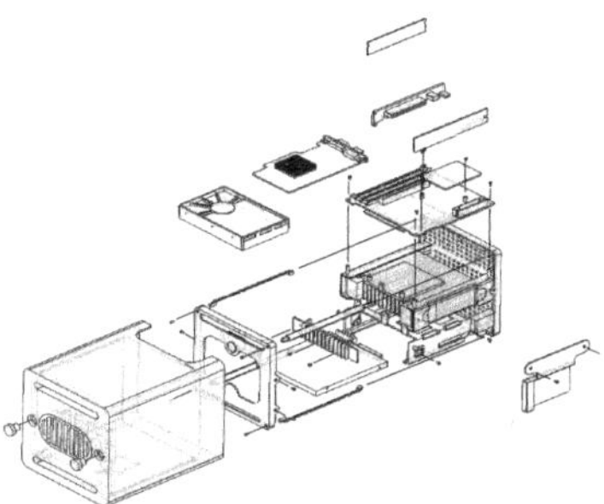

Figure 2-10. Assembly of Power Mac G4 Cube$^{®}$.

Table 2-2 shows the material composition m_{ij} of components A–J in Fig. 2-11b. For components C–F, the material composition data in (Goosey and Kellner, 2003) is utilized. Table 2-3 shows energy intensity me_j^{intens}, recovered energy me_j^{recove}, and material values $mr_j^{recycle}$ (Kuehr and Williams, 2003; Hula, 2003). Considering Apple Computer's Electronic Recycling Program in United States and Canada (www.apple.com/environment), the EOL Power Mac G4 Cubes$^{®}$ are assumed to be transported to one of two facilities in United States (Worcester, MA and Gilroy, CA) for reuse, recycle, and landfill. The average distance between the collection point and the facility is estimated as $D_i = 1000\,\mathrm{km}$ for all components. It is assumed that 40 ton tracks are used for transportation. Based on this assumption, Table 2-4 shows the revenues, costs and energy consumptions of components A–J calculated using Eqs. 18–27. Revenue from reuse r_i^{reuse} reflects current values in the PC reuse markets in the United States (www.dvwarehouse.com and store.yahoo.com/hardcoremac/hardware.htm). Note that reuse option is not available to components A (frame) and B (heat sink).

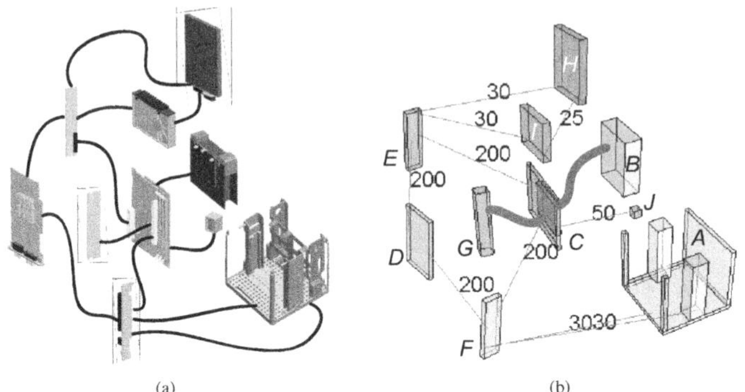

Figure 2-11.   (a) Ten major components and their primary liaisons, and (b) contact and distance specifications.

Table 2-1.   Relative manufacturing difficulty of the locators in the locator library in Fig. 2-4

Locator	Lug	Track	Catch	Boss	Screw	Slot
Mfg. difficulty	20	30	10	70	20	20

Table 2-2.   Material composition [kg] of components A–J in Fig. 2-11.

Component	Aluminum	Steel	Cupper	Gold	Silver
A (frame)	1.2	0	0	0	0
B (heat sink)	0.6	0	0	0	0
C (circuit board)	1.5e-2	0	4.8e-2	7.5e-5	3.0e-4
D (circuit board)	1.0e-2	0	3.2e-2	5.0e-5	2.0e-4
E (circuit board)	4.0e-3	0	1.3e-2	2.0e-5	8.0e-5
F (circuit board)	5.0e-3	0	1.6e-2	2.5e-5	1.0e-4
G (memory)	2.0e-3	0	6.4e-3	2.0e-5	4.0e-5
H (CD drive)	0.25	0.25	0	0	0
I (HD drive)	0.10	0.36	6.4e-3	1.0e-5	4.0e-5
J (battery)	8.0e-5	0	1.4e-3	0	0

Component	Tin	Lead	Cobalt	Lithium	Total
A (frame)	0	0	0	0	1.2
B (heat sink)	0	0	0	0	0.60
C (circuit board)	9.0e-3	6.0e-3	0	0	0.30
D (circuit board)	6.0e-3	4.0e-3	0	0	0.20

E (circuit board)	2.4e-3	1.6e-3	0	0	8.0e-2
F (circuit board)	3.0e-3	2.0e-3	0	0	0.10
G (memory)	1.2e-3	8.0e-4	0	0	4.0e-2
H (CD drive)	0	0	0	0	0.50
I (HD drive)	1.2e-3	8.0e-4	0	0	0.50
J (battery)	0	0	3.3e-3	4.0e-3	2.0e-3

Table 2-3. Material information (Kuehr and Williams, 2003; Hula et al., 2003). Underlined values are estimations due to the lack of published data.

Material	me_j^{intens} [MJ/kg]	me_j^{recove} [MJ/kg]	$mr_j^{recycle}$ [$/kg]
Aluminum	2.1e2	1.4e2	0.98
Steel	59	19	0.22
Cupper	94	85	1.2
Gold	8.4e4	7.5e4	1.7e4
Silver	1.6e3	1.4e3	2.7e2
Tin	2.3e2	2.0e2	6.2
Lead	54	48	1.0
Cobalt	8.0e4	6.0e4	38
Lithium	1.5e3	1.0e3	7.5

Table 2-4. revenue (r [$]), cost ($c$ [$]) and energy consumption (e [MJ]) of the major components A–J.

	A	B	C	D	E
r_i^{reuse}	N/A	N/A	3.5e2	80	1.3e2
$r_i^{recycle}$	1.2	0.60	1.5	1.0	0.39
c_i^{trans}	0.25	0.12	6.2e-2	4.1e-2	1.7e-2
c_i^{refurb}	N/A	N/A	1.8e2	40	65
c_i^{shred}	0.14	7.2e-2	3.6e-2	2.4e-2	9.6e-3
$c_i^{landfill}$	2.4e-2	1.2e-2	6.0e-3	4.0e-3	1.6e-3
e_i^{reuse}	−2.6e2	−1.3e2	−17	−12	−4.5
e_i^{trans}	1.4	0.70	0.35	0.23	9.4e-2
e_i^{refurb}	2.7	1.3	0.66	0.44	0.18
e_i^{rcycle}	−170	−84	−14	−9.5	−3.8
e_i^{shred}	1.2	0.60	0.30	0.20	8.0e-2
$e_i^{landfill}$	2.4e4	1.2e4	6.0e3	4.0e3	1.6e3

	F	G	H	I	J
r_i^{reuse}	39	57	40	60	5.0

$r_i^{recycle}$	0.49	0.36	0.30	0.37	0.12
c_i^{trans}	2.1e-2	8.3e-3	0.10	0.10	4.1e-3
c_i^{refurb}	20	29	20	30	2.5
c_i^{shred}	1.2e-2	4.8e-3	6.0e-2	6.0e-2	2.4e-3
$c_i^{landfill}$	2.0e-3	8.0e-4	1.0e-2	1.0e-2	4.0e-4
e_i^{reuse}	−5.6	−3.1	−68	−45	−2.6e2
e_i^{trans}	0.12	4.7e-2	0.59	0.59	2.3e-2
e_i^{refurb}	0.22	8.8e-2	1.1	1.1	4.4e-2
$e_i^{recycle}$	−4.8	−2.7	−40	−23	−2.0e2
e_i^{shred}	0.10	4.0e-2	0.50	0.50	2.0e-2
$e_i^{landfill}$	2.0e3	8.0e2	1.0e4	1.0e4	4.0e2

4.2   Results

After running the multi-objective genetic algorithm for approximately 240 hours (10 days) on a desktop PC with a 3.2 GHz CUP and a 2 GB RAM (number of population and generation are 100 and 300), thirty seven (37) Pareto optimal designs were obtained as design alternatives. Since the number of objective functions is four, the resulting 4-dimensional space is projected on to six 2-dimensional spaces in Fig. 2-12a–f. Fig. 2-13 shows five representative designs R_1, R_2, R_3, R_4 and R_5. Their objective function values are listed in Table 2-5 and also plotted on a bar chart in Fig. 2-14. As seen in Fig. 2-12, designs R_1, R_2, R_3, and R_4 are the best results only considering an objective function f_1, f_2, f_3 and f_4 regardless of the other objective function values, whereas R_5 is a balanced result in all four objectives.

The spatial configurations of R_3 and R_5 are quite similar, with noticeable differences in the EOL treatments. Figs. 2-15 and 2-16 show one of the optimal disassembly sequences of R_3 and R_5 with the EOL treatments of components, respectively. Design R_3 (design biased for profit) uses three screws, one of which is used between components A and B. Since components A and B have no reuse options, and recycling them is less economical than land-filling due to high labor cost for removing screws; they are not disassembled and simply discarded altogether for higher profit. On the other hand, components A and B are disassembled and recycled in R_5 (balanced design for all objectives) to reduce environmental impact at the expense of higher disassembly cost (lower profit).

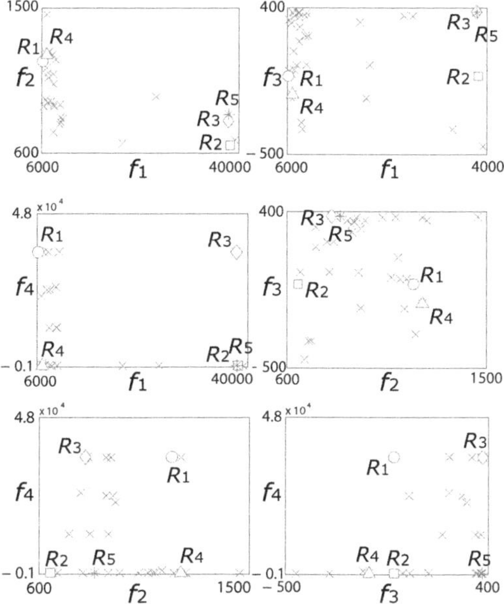

Figure 2-12. Distribution of Pareto optimal designs in six 2-dimensional spaces (a)–(f).

Table 2-5. Objective function values of R_1, R_2, R_3, R_4 and R_5

	F_1 (dist. spec.)	f_2 (mfg. diff.)	f_3 (profit)	f_4 (env. impact)
R_1	6175	1170	-19.30	35627
R_2	38496	650	-19.34	-642
R_3	38227	800	374.72	35593
R_4	6884	1210	-130.79	-741
R_5	38299	840	373.24	-647

As stated in the previous section, reuse, if available, is usually the best EOL treatment for a component because of its high revenue and high energy recovery. For the components without reuse option, the choice between recycle and landfill depends on the ease of disassembly, as seen in these results. If the disassembly cost is low enough that recycling the component is more profitable than land-filling it, recycle becomes the most profitable EOL treatment. Otherwise, there is a trade-off between the profit and the environmental impact, which is found in the Pareto optimal designs.

Oftentimes such trade-off among alternative designs can hint at opportunities for further design improvements. For example, the examination of the differences between R_3 and R_5 suggests the possibility of replacing the screws between A and B by slot-like locators (which are not available for A and B in the locator library) for higher profit and lower environmental impact.

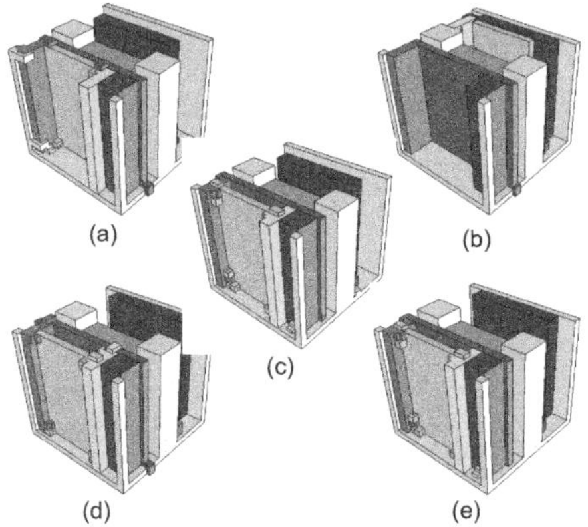

Figure 2-13.   Representative Pareto designs: (a) R_1, (b) R_2, (c) R_3, (d) R_4 and (e) R_5.

5.   SUMMARY AND FUTURE WORKS

This paper presented an extension of our previous work on a computational method for product-embedded disassembly, which newly incorporates EOL treatments of disassembled components and subassemblies as additional decision variables, and LCA focusing on EOL treatments as a means to evaluate environmental impacts. The method was successfully applied to a realistic example of a desktop computer assembly, and a set of Pareto optimal solutions is obtained as design alternatives.

Future work includes the adoption of more detailed LCA covering entire product life including the production and use phases, the development of more efficient optimization algorithm, the study on the effect of embedded disassembly on assembly, and the derivation of the generalizable design rules through the comparison of the optimization results with the existing designs of other product types.

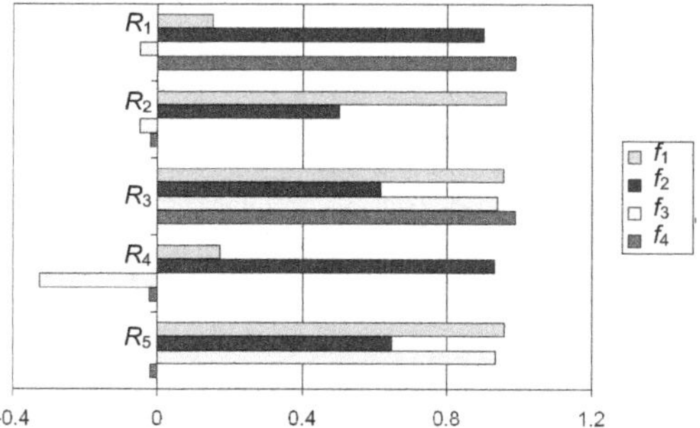

Figure 2-14. Objective function values of R1, R2, R3, R4 and R5 (scaled as f1 : 1/40000, f2 : 1/1300, f3 : 1/400, f4 : 1/36000).

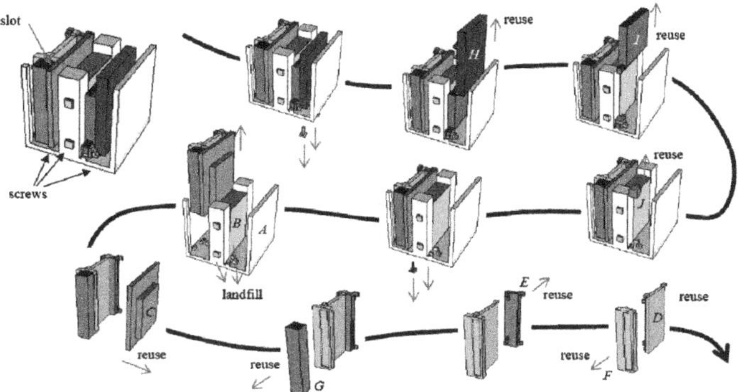

Figure 2-15. Optimal disassembly sequence of R_3 with the optimal EOL treatments of components.

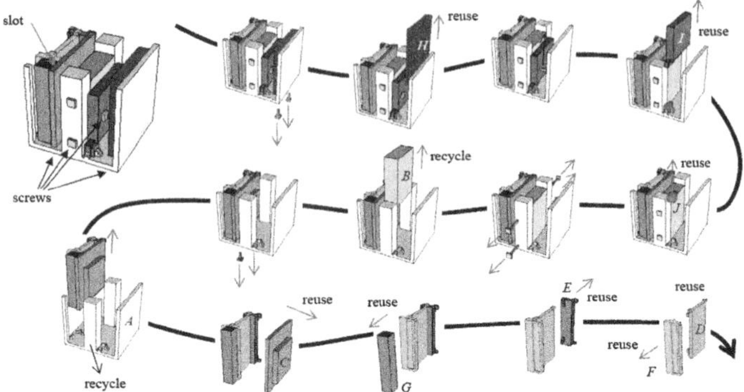

Figure 2-16. Optimal disassembly sequence of R_5 with the optimal EOL treatments of components.

Acknowledgments

The funding for this research was provided by the National Science Foundation of the United States through grant # BES-0124415. Any options, findings, and conclusions or recommendations expressed in this material are those of the authors and do not necessarily reflect the views of the National Science Foundation.

References

Aanstoos, T.A., Torres, V.M., and Nichols, S.P. (1998) Energy model for end-of-life computer disposition, *IEEE Transactions on components, packaging, and manufacturing technology*, 21(4): 295–301.

Baldwin, D.F., Abell, TE., Lui, M.-C., De Fazio, T.L., and Whitney, D.E. (1992) An integrated computer aid for generating and evaluating assembly sequences for mechanical products, *IEEE Transactions on Robotics and Automation*, 7(1): 78–94.

Beasley, D. and Martin, R.R. (1993) Disassembly sequences for objects built from unit cubes, *Journal of Compute-Aided Design*, 25(12): 751–761.

Bonenberger, P.R. (2000) *The First Snap-Fit Handbook: Creating Attachments for Plastic Parts*, Hanser Gardner Publications, München, Germany.

Boothroyd, G. and Alting, L. (1992) Design for assembly and disassembly, *Annals of CIRP*, 41(22): 625–636.

Boothroyd, G., Dewhurst, P., and Knight, W. (1994) *Product Design for Manufacture and Assembly*, Marcel Dekker, Inc., New York, NY.

Caudill, J.R., Zhou, M., Yan, P., and Jim, J. (2002) Multi-life cycle assessment: an extension of traditional life cycle assessment, In: M.S. Hundal (ed.), *Mechanical Life Cycle Handbook*, Marcel Dekker. New York, NY, pages 43–80.

Chen, R.W., Navinchandra, D., and Prinz, F. (1993) Product design for recyclability: a cost benefit analysis model and its application, *IEEE Transactions on Components, Packaging, and Manufacturing Technology*, 17(4): 502–507.

Chen, S.-F., Oliver, J.H., Chou, S.-Y., and Chen, L.-L. (1997) Parallel disassembly by onion peeling, *Transactions of ASME, Journal of Mechanical Design*, 119(22): 267–274.

Corcoran III, A.L. and Wainwright, R.L. (1992) A genetic algorithm for packing in three dimensions, Proceedings of *the ACM/SIGAPP Symposium on Applied Computing*, Kansas City, Missouri, pages 1021–1030.

Das, S.K., Yedlarajiah, P., and Narendra, R. (2000) An approach for estimating the end-of-life product disassembly effort and cost, *International Journal of Production Research*, 38(3): 657–673.

De Fazio, T.L. and Whitney, D.E. (1987) Simplified generation of all mechanical assembly, *IEEE Transactions of Robotics and Automation*, 3(6): 640–658.

Deb, K., Pratap, A., Agarwal, S., and Meyarivan, T. (2002) A fast and elitist multiobjective genetic algorithm: NSGA-II, *IEEE Transactions on Evolutionary Computation*, 6(2): 182–197.

Desai, A. and Mital, A. (2003) Evaluation of disassemblability to enable design for disassembly in mass production, *International Journal of Industrial Ergonomics*, 32(4): 265–281.

Dutta, D. and Woo, T.C. (1995) Algorithm for multiple disassembly and parallel assemblies, *Transactions of ASME, Journal of Engineering for Industry*, 117: 102–109.

Fonseca, C.M. and Fleming, P.J. (1993) Genetic algorithms for multiobjective optimization: formulation, discussion and generalization, Proceedings of *the 5th International Conference on Genetic Algorithms*, July 17–22, Urbana-Champaign, IL, pages 416–423.

Fujita, K., Akagi, S., and Shimazaki, S. (1996) Optimal space partitioning method based on rectangular duals of planar graphs, *JSME International Journal*, 60: 3662–3669.

Glover, F. (1974) *Heuristics for Integer Programming using Surrogate Constraints*, Business Research Division, University of Colorado.

Glover, F. (1986) Further paths for integer programming and links to artificial intelligence, *Journal of Computer and Operations Research*, 13(5): 533–549.

Goggin, K. and Browne, J. (2000) The resource recovery level decision for end-of-life products, *Production Planning and Control*, 11(7): 628–640.

Goosey, M. and Kellner, R. (2003) Recycling technologies for the treatment of end of life printed circuit boards (PCBs), *Circuit World*, 29(3): 33–37.

Grignon, P.M. and Fadel, G.M. (1999) Configuration design optimization method, Proceedings of *the ASME Design Engineering Technical Conferences and Computers in Engineering Conference*, September 12–15, Las Vegas, Nevada, DETC99/DAC-8575.

Homem dé Mello, L.S. and Sanderson, A.C. (1990) AND/OR graph representation of assembly plans, *IEEE Transactions on Robotics and Automation*, 6(2): 188–199.

Homem dé Mello, L.S. and Sanderson, A.C. (1991) A correct and complete algorithm for generation of mechanical assembly sequences, *IEEE Transactions on Robotics and Automation*, 7(2): 228–240.

Hula, A., Jalali, K., Hamza, K., Skerlos, S., and Saitou, K. (2003) Multi-criteria decision making for optimization of product disassembly under multiple situations, *Environmental Science and Technology*, 37(23): 5303–5313.

Jain, S. and Gea, H.C. (1998) Two-dimensional packing problems using genetic algorithm, *Journal of Engineering with Computers*, 14: 206–213.

Kaufman, S.G., Wilson, R.H., Jones, R.E., Calton, T.L., and Ames, A.L. (1996) The Archimedes 2 mechanical assembly planning system, Proceedings of *the IEEE International Conference on Robotics and Automation*, April, 1996, Minneapolis, Minnesota, pages 3361–3368.

Kroll, E., Beardsley, B., and Parulian, A. (1996) A methodology to evaluate ease of disassembly for product recycling, *IIE Transactions*, 28(10): 837–845.

Kolli, A., Cagan, J., and Rutenbar, R. (1996) Packing of generic, three-dimensional components based on multi-resolution modeling, Proceedings of *the ASME Design Engineering Technical Conferences and Computers in Engineering Conference*, August 18–22, Irvine, California, DETC/DAC-1479.

Kuehr, R. and Williams, E. (Eds.) (2003) *Computers and the Environment*, Kluwer Academic Publishers, Dordrecht, The Netherlands.

Kuo, T. and Hsin-Hung, W. (2005) Fuzzy eco-design product development by using quality function development, Proceedings of the *EcoDesign: Fourth International Symposium on Environmentally Conscious Design and Inverse Manufacturing*, December 12–14, Tokyo, Japan, 2B-3-3F.

Lambert, A.J.D. (1999) Optimal disassembly sequence generation for combined material recycling and part reuse, Proceedings of *the IEEE International Symposium on Assembly and Task Planning*, Portugal, pages 146–151.

Lee, S. and Shin, Y.G. (1990) Assembly planning based on geometric reasoning, *Computer and Graphics*, 14(2): 237–250.

Li, J.R., Tor, S.B., and Khoo, L.P. (2002) A hybrid disassembly sequence planning approach for maintenance, *Transactions of ASME, Journal of Computing and Information Science in Engineering*, 2(1): 28–37.

Matsui, K., Mizuhara, K., Ishii, K., and Catherine, R.M. (1999) Development of products embedded disassembly process based on end-of-life strategies,

Proceedings of *EcoDesign: First International Symposium on Environmentally Conscious Design and Inverse Manufacturing*, February 1–3, Tokyo, Japan, pages 570–575.

Minami, S., Pahng, K.F., Jakiela, M. J., and Srivastave, A. (1995) A cellular automata representation for assembly simulation and sequence generation, Proceedings of the *IEEE International Symposium on Assembly and Task Planning*, August 10–11, Pittsburgh, Pennsylvania, pages 56–65.

O'Shea, B., Kaebernick, H., Grewal, S.S., Perlewitz, H., Müller, K., and Seliger, G. (1999) Method for automatic tool selection for disassembly planning, *Assembly Automation*, 19(1): 47–54.

Reap, J. and Bras, B. (2002) Design for disassembly and the value of robotic semi-destructive disassembly, Proceedings of *the ASME Design Engineering Technical Conferences and Computers and Information in Engineering Conference*, September 29 – October 2, Montreal, Canada, DETC2002/DFM-34181.

Rose, C.M. and Stevels, A.M. (2001) Metrics for end-of-life strategies (ELSEIM), *Proceedings of the IEEE International Symposium on Electronics and the Environment*, May 7–9, Denver, Colorado, pages 100–105.

Sodhi, R., Sonnenberg, M. and Das, S. (2004) Evaluating the unfastening effort in design for disassembly and serviceability, *Journal of Engineering Design*, 15(1): 69–90.

Srinivasan, H. and Gadh, R. (2000) Efficient geometric disassembly of multiple components from an assembly using wave propagation, *Transactions of ASME, Journal of Mechanical Design*, 122(2): 179–184.

Sung, R.C.W., Corney, J.R., and Clark, D.E.R. (2001) Automatic assembly feature recognition and disassembly sequence generation, *Transactions of ASME, Journal of Computing and Information Science in Engineering*, 1(4): 291–299.

Takeuchi, S. and Saitou, K. (2005) Design for product-embedded disassembly, Proceedings of *the ASME Design Engineering Technical Conferences*, Long Beach, California, September 24–28, DETC2005-85260.

Takeuchi, S. and Saitou, K. (2006) Design for optimal end-of-life scenario via product-embedded disassembly, Proceedings of *the ASME Design Engineering Technical Conferences*, Philadelphia, Pennsylvania, September 10–13, DETC2006-99475.

Williams, E.D. and Sasaki, Y. (2003) Energy analysis of end-of-life options for personal computers: resell, upgrade, recycle, Proceedings of the *IEEE International Symposium on Electronics and the Environment*, May 19–22, Boston, MA, pages 187–192.

Woo, T.C. and Dutta, D. (1991) Automatic disassembly and total ordering in three dimensions, *Transactions of ASME, Journal of Engineering for Industry*, 113: 207–213.

Chapter 3

MULTI-LEVEL DECOMPOSITION FOR TRACTABILITY IN STRUCTURAL DESIGN OPTIMIZATION

Erik D. Goodman[1,2,3], Ronald C. Averill[1,3] and Ranny Sidhu[1]

[1] *Red Cedar Technology, 4572 S. Hagadorn Rd, Suite 3-A, East Lansing, Michigan 48823;*
[2] *Dept. of Electrical and Computer Engineering; and* [3] *Department of Mechanical Engineering, Michigan State University*

Abstract   This paper describes two approaches that allow decomposition of structural design problems that enable structural design optimization to be performed on systems that are often seen as too large or complex to address in a single optimization. The COMPOSE method is shown to enable optimization of many tightly coupled subsystems despite the usual problem of non-convergence when a subsystem is repeatedly removed and optimized under the previously prevailing boundary conditions, then reinserted into the entire system model, which subsequently causes the boundary conditions experienced by the subsystem to change. Use of COMPOSE may allow reducing the number of whole-system evaluations necessary for component design from hundreds or thousands to fewer than ten, while still exploring the component design space extensively. A second approach, collaborative independent agents, is shown to address problems that have both large design spaces and time-intensive analyses, rendering them intractable to traditional methods. In an example problem, a set of loosely coupled optimization agents is shown to reduce dramatically the computing time needed to find good solutions to such problems. The savings result from the continuing transfer of results from rapid, low-refinement, less accurate search to agents that search at the full level of refinement and accuracy demanded of a solution of the problem, essentially providing them guidance as to what portions of the design space are likely to be worth searching in greater detail.

Keywords: structural decomposition, multi-level decomposition, COMPOSE, structural design automation, heterogeneous agents, multi-agent optimization, collaborative independent agents.

Erik D. Goodman et al.: *Multi-Level Decomposition for Tractability in Structural Design Optimization*, Studies in Computational Intelligence (SCI) **88**, 41–62 (2008)
www.springerlink.com

1. INTRODUCTION

In large complex engineered systems, often only a subsystem or a small part of the system design needs to be modified to adapt or improve performance in some way. For example, to improve frontal crash safety in an automobile, an engineer might focus design changes on only the vehicle lower front compartment rails and the bumper. Or, an aerospace engineer may focus on a single spar within a structure that is behaving in a nonlinear manner. Depending on the degree to which the performance of the subsystem or component being designed influences the behavior of the whole system of which it is a part, the design of the component may or may not need to be performed in the context of the whole system — i.e., involve simulation of the whole system to assure that the boundary conditions presented to the component are representative of those to be encountered in service. Two approaches to addressing such problems are presented here: the first (called COMPOSE) is for problems in which that coupling is relatively strong, and the second, collaborative independent agents, is useful for problems where the component can be designed in relative isolation, or can be coupled with COMPOSE in problems where the interaction of component and system must be represented dynamically during the optimization process.

1.1 Tightly Coupled Subsystems

In designing structures in which the subsystem behavior is strongly coupled to that of the overall system in such a way that even small changes to the subsystem can strongly affect the *interactions* between the system and subsystem, design optimization of the subsystem usually requires that a mathematical/computational model of the complete system be used so that these interactions can be taken into account directly. These full system level models are often very large and complicated, and thus a significant amount of CPU time (e.g., 10–30 hours) is required to simulate the performance of each new design scenario. Because many thousands of design evaluations might be necessary to perform a high fidelity design optimization involving 10–250 design variables, it might take weeks or months, even using multiple machines, to perform a design optimization study on even a small subsystem.

The COMPOSE methodology has been introduced to drastically reduce the time and effort required to perform design optimization on subsystems whose performance is strongly coupled to that of the complete system to which they belong. This approach commonly reduces the CPU time for such design studies by a factor of 10^1–10^3, depending upon the problem definition. Further, this approach has the potential to yield better results than are attainable by directly applying most optimization methods to a full system model, because a much more thorough search is performed. Robustness of the design can also be enforced with this approach. COMPOSE, an acronym for COMPonent Optimization within a System Environment, is aimed at enabling engineers to find, in much less time, designs that, while not necessarily globally optimal, exhibit significantly higher performance and robustness than are achieved with traditional optimization methods.

This chapter will describe not only the basic COMPOSE methodology, but also an extension in which the system S is hierarchically decomposed into N subsystems, wherein the ith subsystem is further decomposed into N_i subsystems, each of which is further decomposed into N_{ij} subsystems, and so on, as shown in Figure 3-1. This approach significantly enhances the speed of optimization by allowing different numbers and types of design variables to be studied at the various hierarchical levels. An example of application of COMPOSE in automotive chassis design is given.

1.2   Quasi-Independent Subsystems

In other cases, the boundary conditions of interest in determining a component's behavior have been specified, removing the need to simulate the performance of the component in the context of a larger system in order to evaluate each design candidate. In such cases, the COMPOSE methodology is not needed. However, when the design space is large and the evaluation time per design is also large, the problem may still not be a practical candidate for traditional optimization approaches. The collaborative independent agents methodology is described here as an approach to reducing dramatically the amount of time needed to find excellent design candidates, even allowing for a specified amount of stochastic variability in design variables and loading conditions. An example of use of collaborative independent agents in automotive lower compartment rail design is provided in the second part of this chapter.

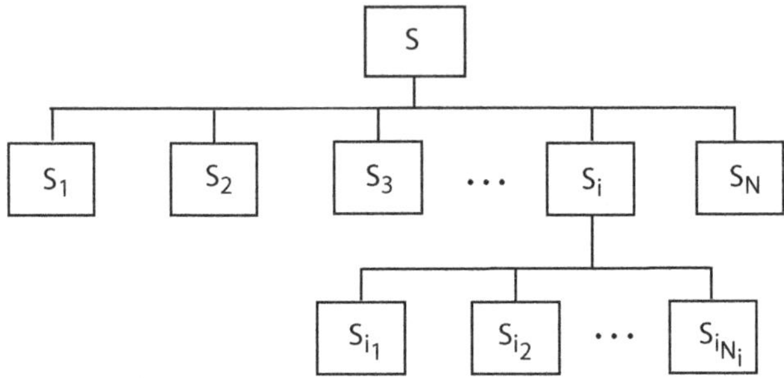

Figure 3-1.   Hierarchical decomposition of the system S.

2.   BACKGROUND

Optimization is a process that seeks to identify the best design according to a specified measure, or objective function, while also satisfying certain restrictions, or constraints. But computational methods used for optimization may not always identify the best design, and in these cases we seek to find as good a design as possible, or at least a design that is better than the existing one, within a reasonable timeframe and cost. Herein, use of the word optimal is intended to mean best, or as good as possible, or better than before, depending upon what is practical in that context and application. Also, in real-world situations, we often seek a design that is optimal with respect to performance, robustness and reliability rather than one exhibiting absolute peak performance under a very tightly specified set of deterministic environmental conditions.

Gradient-based optimization techniques have been successfully applied to many structural optimization problems, e.g., (Soto and Diaz, 1993; Suzuki and Kikuchi, 1990; Suzuki and Kikuchi, 1991). However, these methods have several drawbacks. First, they tend to find quickly and get stuck on local extrema (Suzuki and Kikuchi, 1991). In addition, gradient methods are not suitable for finding singular extrema, or for optimizing problems with discontinuous or noisy response functions (e.g., crash problems).

Optimization methods based on genetic algorithms (GAs) and other evolutionary methods have been applied to various engineering problems, e.g., (Averill et al., 2001; Averill et al., 1995; Chapman et al., 1993; Daratech, 2002; Eby et al., 1999; Eby et al., 1997; Eby et al., 1998; Leung and Nevill, 1994; LeRiche and Haftka, 1993; Lin et al., 1994; Mallott et al., 1996; Nagendra et al., 1992; Nagendra et al., 1993; Red Cedar Technology, 2001; Sandgren et al., 1990), and have demonstrated the potential to overcome many of the problems associated with gradient-based methods. However, the need of GAs to evaluate many alternative designs often limits them to use only in problems in which the design space can be made sufficiently small, even though GAs are most effective (relatively) when the design space is large.

The concept of multi-level solution of design optimization problems has been investigated for more than 20 years—see, for example, (Haftka, 1984; Papadrakakis and Tsompanakis, 1999; Sobieszczanski-Sobieski et al., 2003; Sobieszczanski-Sobieski et al., 1985; Sobieszczanski-Sobieski et al., 1987). Yet its applications have often been limited by the specific nature of the algorithm, or by the requirement to calculate sensitivity derivatives of the subsystem parameters with respect to the system level response. In the first method described here, some shortcomings of existing multi-level decomposition methods are overcome in a fairly general way, providing a robust method for solving several very wide classes of multi-level design optimization problems.

We are concerned here with large optimization problems, where large may refer to the number of design variables and/or the CPU time required to evaluate the objective function for a particular design candidate. In such cases it is common to break the problem into smaller parts, or *subsystems*, using *decomposition*.

Decomposition may be applied to the optimization problem itself, to the physical/temporal domain of the system, and/or to the disciplinary responses in a multidisciplinary problem. The discussion immediately following focuses primarily on spatial decomposition, wherein the physical system is decomposed into several subsystems. The COMPOSE algorithm is not limited to such problems, however.

After an optimization problem is decomposed, the solution procedure may take one of several forms. Among the more popular methods is a multi-level optimization procedure. For example, in a two-level optimization procedure the optimization of the subsystem variables, $\mathbf{x}^i$, is nested inside an upper-level optimization of the global variables, $\mathbf{z}$. It is also possible to define a third set of variables, $\mathbf{y}$, that are output from one subsystem and input to another subsystem (Sobieszczanski-Sobieski et al., 2003). An iterative approach can then be used to coordinate the identification of the subsystem and global variables that jointly optimize the system. Most such iterative approaches depend upon the calculation of sensitivity derivatives of the optima of each subsystem with respect to changes in the global variables, $\mathbf{z}$. Often, the calculation of these derivatives is either very difficult or computationally very expensive. In some cases, the sensitivity derivatives are discontinuous. The cost of calculating the sensitivity derivatives depends in part on the front of interaction between the subsystems and the number of design variables.

A solution procedure that does not require the calculation of sensitivity derivatives would be beneficial in many applications. Such an approach is often called direct iteration, or fixed-point iteration. This technique, however, has less than desirable convergence characteristics when applied to some classes of problems. Namely, problems in which large changes in the interaction variables occur during the iteration process may not converge to a near optimal solution, and may fail to converge at all. The COMPOSE algorithm uses a fixed-point iteration algorithm that significantly enhances its ability to converge when applied to multilevel optimization of large problems.

3. GENERAL PROBLEM STATEMENT—COMPOSE PROBLEMS

Consider any continuous or discrete system that exists in the domain Ω, as shown in Figure 3-2(a). The spatial and temporal performance of the system under a prescribed set of environmental conditions (generalized loads) can be described mathematically by equations (e.g., differential, integral, algebraic, etc.) in terms of primary variable(s) denoted *u(x,y,z,t)* and the secondary variable(s) denoted *f(x,y,z,t)*. The boundary of Ω is denoted as Γ.

Within the system domain Ω, one or more subsystems $\Omega_{subsystem(i)}$ $(i = 1$ to $N)$ may be identified, as shown in Figure 3.2(b). The only restriction on the definition of these subsystems is that their domains may not overlap. Subsystems i and j must not have any common interior points for all $i, j = 1$ to N, but subsystems may have common boundary points. The subsystem boundary $\Gamma_{subsystem(i)}$ represents the boundary between subsystem i and the remainder of the system, as shown in Figure 3.2(c). We assume here without loss of generality that:

$$u = \hat{u}_{subsystem(i)} \; on \; \Gamma_{subsystem(i)} \tag{1}$$

Of interest here is the common situation in which the performance of one or more subsystem designs is to be optimized by changing one or more characteristics (design variables) of the subsystem(s). The subsystems at a given level do not share any design variables, and the remainder of the system is fixed so there are no global design variables. In this context, a subsystem is optimized when a specified objective function is minimized or maximized, including the special case in which the subsystem satisfies a particular performance target. The subsystem designs may also be subject to a set of constraints that must be satisfied. The optimization is performed by finding the simultaneous values of a set of design variables that extremize the objective function while satisfying all constraints. Mathematically, the optimization statement within each subsystem may take the form:

$$\text{Minimize (or maximize): } F_i(x_1, x_2, \ldots, x_n)^i$$
$$\text{such that: } G_{ij}(x_1, x_2, \ldots, x_n)^i < 0, \; j = 1, 2, \ldots, p_i$$
$$H_{ij}(x_1, x_2, \ldots, x_n)^i = 0, \; j = 1, 2, \ldots, q_i \tag{2}$$

where $(x_1, x_2, \ldots, x_n)^i$ are design variables in subsystem i

- $F_i(x_1, x_2, \ldots, x_n)^I$ is the objective (performance) function in subsystem i
- $G_{ij}(x_1, x_2, \ldots, x_n)^i$ are inequality constraints in subsystem i
- $H_{ij}(x_1, x_2, \ldots, x_n)^i$ are equality constraints in subsystem i

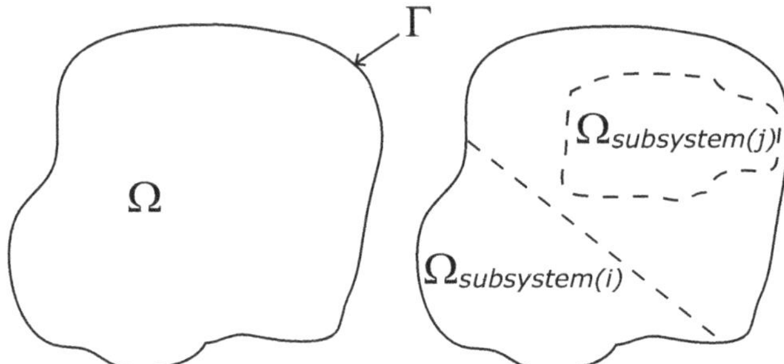

Figure 3-2.   Schematic of system and subsystem domains and their boundaries.

The problem statement in Eq. (2) is intended to include optimization problem statements in the broadest sense, including multi-objective optimization.

In some cases, even major changes to a subsystem design do not strongly affect the *interactions* between the system and the subsystem(s). In other words, there are some systems in which the subsystem essential boundary conditions $\hat{u}_{subsystem(i)}$ experience small or no change when the values of design variables in any subsystem are modified. In these cases, the subsystem i can be redesigned in isolation using mathematical models involving only the domain $\Omega_{subsystem(i)}$, which should be smaller and simpler than that of the entire system. The system contributions are included through the boundary conditions $\hat{u}_{subsystem(i)}$.

Here we will treat the cases in which the subsystem behavior is strongly coupled to that of the overall system in such a way that even moderate changes to a subsystem can strongly affect the interactions between the system and subsystem(s). The objective for such problems is to drastically reduce the time and effort required to perform design optimization on subsystems whose performance is strongly coupled to that of the complete system to which they belong.

Let us assume that a given design optimization statement as in Eq. (2) requires that a minimum number of design evaluations be performed, this number of evaluations depending primarily upon the number of design variables, the nature of the design space, and the optimization search algorithm employed. Then, a reduction in the computational effort required to optimize a subsystem must be achieved by reducing the computational effort to evaluate each design scenario. Here, a technique is sought in which most design evaluations can be performed using the subsystem mathematical models, which should be much smaller and computationally more efficient than the complete system level model. But such an approach must also account for the sometimes strong interactions between the performance of the system and the subsystem(s).

4.   DESCRIPTION OF THE COMPOSE METHOD

In a typical design optimization problem, the goal is to design a system so that it behaves in a prescribed or optimal manner in a given environment or under a set of prescribed conditions. The challenge of the current problem is to simultaneously identify both a subsystem that is optimal according to a specified criterion and the subsystem boundary conditions under which the subsystem should behave optimally.

In the general case, the subsystem boundary conditions associated with the optimal design cannot be known until the design approaches its optimal form, and the final optimal design cannot be identified until the subsystem boundary conditions approach a form associated with the final optimal design. In other words, the optimal design and the subsystem boundary conditions are interdependent, and they must be codetermined.

A direct iterative approach has been devised to solve this problem without the need for calculating sensitivity derivatives. For two levels, the algorithm works as shown in Figure 3-3. Note that the subsystem optimization in step 3 is typically terminated prior to convergence to the optimal solution. Often, there is no point in expending the extra effort toward finding an exact subsystem

optimum prior to identifying subsystem boundary conditions that are close to
their final form. Hence, the iterative process often proceeds using near-optimal
subsystem solutions.

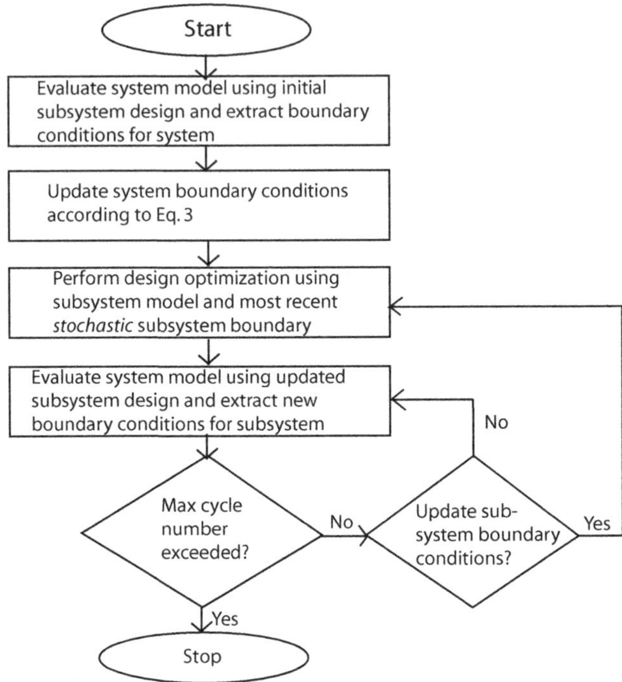

Figure 3-3. Flowchart for the COMPOSE iteration strategy to solve two-level optimization
problem.

It has been found that convergence toward an improved solution is greatly improved when the optimized subsystem design(s) in iteration k:

1. have good performance under the kth set of subsystem boundary conditions; and
2. exhibit similar performance characteristics under the kth and $(k + 1)^{th}$ boundary conditions (i.e., the system and subsystem solutions in steps 3 and 4 (see Figure 3-2), respectively, do not have significantly different gradients or eigenmodes).

Thus, it is important during intermediate iterations to identify optimal or near optimal subsystem designs that have similar performance under small to moderate variations in the subsystem boundary conditions. Optimal subsystem designs that satisfy the above criteria are said to be robust against stochastic variations in the subsystem boundary conditions.

Convergence may also be improved by reducing the magnitude of the change in subsystem boundary conditions from one iteration to the next, or by using a weighted average of the boundary conditions at two consecutive steps.

In order to satisfy the conditions above, the subsystem boundary conditions at iteration k can be cast in the form:

$$\hat{u}_{subsystem(i)} = \sum_{s=1}^{k} \widetilde{w}_s \hat{u}_s \qquad \text{on } \Gamma_{subsystem(i)} \tag{3}$$

where $\hat{u}_{k-1}(x, y, z, t), \hat{u}_k(x, y, z, t)$ are the subsystem boundary conditions in terms of generalized primary variables on $\Gamma_{subsystem(i)}$ at iterations $k - 1$ and k, respectively; and $\widetilde{w}_1(x, y, z, t), \widetilde{w}_2(x, y, z, t)$ are weight functions whose spatial and temporal distributions are predetermined and whose magnitudes are varied stochastically within a selected range. A common form of Equation (3) is:

$$\hat{u}_{subsystem(i)} = \widetilde{w}_1 \hat{u}_{k-1} + \widetilde{w}_2 \hat{u}_k \qquad \text{on } \Gamma_{subsystem(i)} \tag{4}$$

where $\hat{u}_{k-1}(x, y, z, t), \hat{u}_k(x, y, z, t)$ are the subsystem boundary conditions in terms of generalized primary variables on $\Gamma_{subsystem(i)}$ at iterations $k - 1$ and k, respectively; and $\widetilde{w}_1(x, y, z, t), \widetilde{w}_2(x, y, z, t)$ are weight functions whose spatial and temporal distributions are predetermined and whose magnitudes are varied stochastically within a selected range.

It is possible for the interaction between the system and a subsystem to be specified or constrained along a portion of the boundary $\Gamma_{subsystem(i)}$, whenever this interaction is either known or desired to be of a particular form.

5. MULTIPLE LEVELS OF COMPOSE

While Figure 3-3 describes the process for two-level optimization, recent efforts have been aimed at taking advantage of even higher levels of decomposition to optimize complex problems. For example, consider the three-level optimization of a frame-type body structure for crashworthiness, durability, or other local performance measures. Shape optimization of several members (each denoted S_{ij} in Figure 3-1) may be performed separately and simultaneously at the lowest level in order to identify the most mass efficient shapes for these members. Their gage thicknesses may then be optimized together at the next highest sub-assembly level (denoted S_i in Figure 3-1), while the interactions between the subsystem(s) and the entire system assembly (S) are maintained. In this way, many optimization problems are solved simultaneously using inexpensive sub-system evaluations, the results of which cascade upward into the entire system. Meanwhile, the coupling of the many sub-systems is maintained by cascading these interactions downward from the system level to the various sub-system levels. An iterative approach results in which convergence is encouraged partly by promoting robustness of each subsystem against changes in the interactions among the subsystems.

6. APPLICATION OF MULTI-LEVEL COMPOSE METHOD

The system (truck) and subsystem (lower compartment rails) shown in Figure 3-4 were selected for demonstrating the application of COMPOSE to crashworthiness design problems.

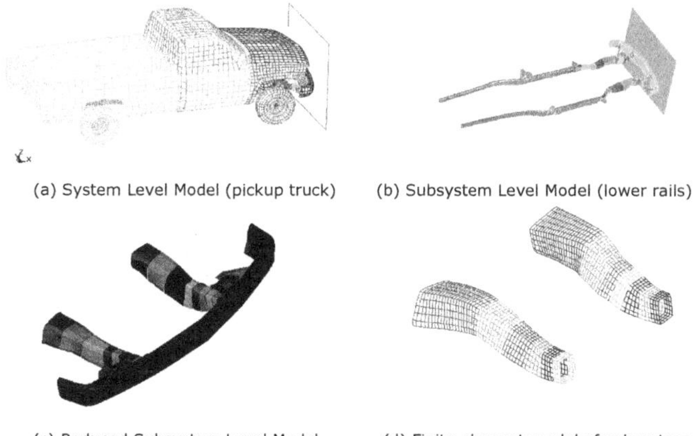

Figure 3-4.　System and subsystem models used for crashworthiness optimization of lower rails.

In this problem, the vehicle impacts a rigid wall with an initial velocity of 35 mph. The shape of the lower compartment rails was designed using 140 shape design parameters, 70 in each rail. The design parameters were actually spline points which determined the cross-sectional shape of the rail at various stations along its length, as shown in Figures. 3-4(c,d) and 3-5. An automatic mesher was used to generate a new mesh for each potential design. Each rail was designed separately. The system- and subsystem-level finite element analyses were performed using LS-DYNA, an explicit finite element code. Boundary conditions at the system/subsystem interface were extracted from the system-level model and then imposed on the subsystem-level model. The automated design optimization was executed on a personal computer during a period of about five days. COMPOSE was implemented within the proprietary design optimization software, HEEDS, developed by the authors' company.

The energy absorbed in the subsystem (rails) was increased by approximately 30% (see Figure 3-6(a)), while the overall energy absorbed by the system (truck) was increased by more than 5.5% (see Figure 3-6(b)). In Figure 3-6(a), the curve denoted "Local EA" represents the increase in energy absorbed in the rails as measured by the subsystem model, while the curve denoted "Global EA" represents the increase in energy absorbed in the rails as measured by the system level model. These results differ slightly due to the multiple contact conditions that occur in the system level model, which change as the rail design is modified. Only six system level evaluations were performed, but complete coupling between the system and subsystem was maintained. This application clearly demonstrates the potential of COMPOSE to solve crashworthiness problems and other classes of design problems that were formerly considered intractable.

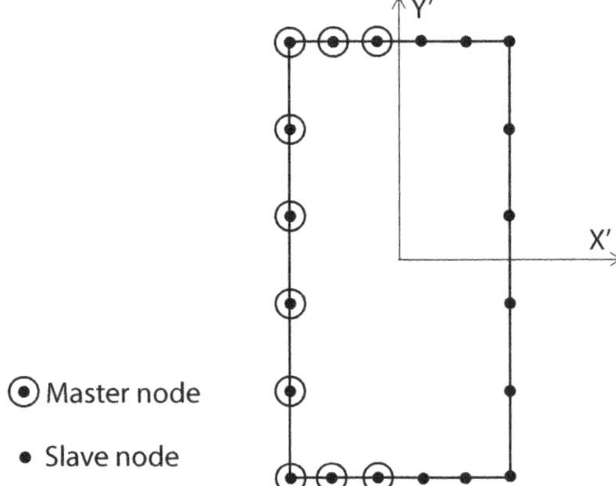

Figure 3-5.   Cross-sectional shape representation. Each node represents a spline point that can move normal to the original rectangular shape in the X′Y′ plane.

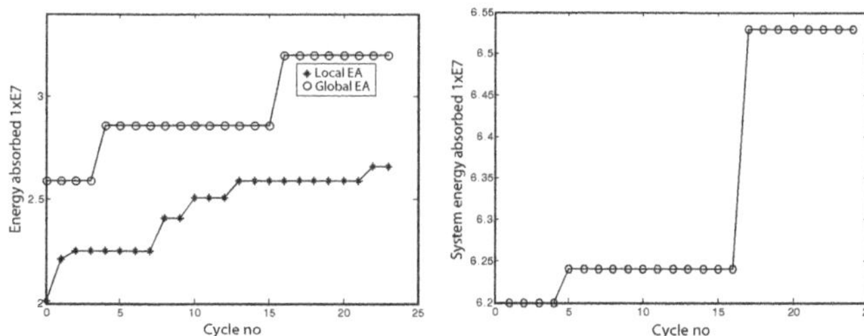

Figure 3-6.   Energy absorbed in subsystem (a) and system (b) as a function of design cycle.

7.   COLLABORATIVE INDEPENDENT AGENTS APPROACH

While the COMPOSE method is critical to "in-situ" optimization of components or subsystems in a system that provides boundary conditions that are strongly coupled to the performance of the component being optimized, there are other situations, or stages in the product design process, in which the component's boundary conditions may be treated as given and relatively insensitive to the component's performance. Furthermore, at the lowest level in a multi-level COMPOSE decomposition of a system, the combination of design space and analysis time per design may still be so large that traditional optimization methods do not allow identification of high-performance designs within the time available. In such cases, the use of collaborative independent agents, as explained next, even within the COMPOSE process, can make an intractable problem into one that can readily be addressed without massively parallel computing facilities and enormous numbers of CAE software licenses.

Collaborative independent agents are quasi-independent search agents running simultaneously in a coordinated search to find the solution to a single problem. They are quasi-independent in the sense that they run asynchronously, on either the same or different processors, with only infrequent, unsynchronized communication amongst them. The agents are typically organized in a hierarchical manner, and they communicate with one another by passing designs from less refined agents to more refined agents. Typically, one or more agents are searching at the level of refinement at which the ultimate design is sought, using the CAE analysis tools at the level of refinement at which the user has full confidence in their capability to determine the suitability of a design for solving the problem. Other agents are searching in simplified design spaces, or with less refined tools, or using coarser approximations, or under only a subset of the loading cases, or with less consideration of stochasticity, etc. Such simplifications are undertaken so that each evaluation takes much less time to compute than at the final target level of resolution. While all agents are performing search operations, the less refined agents are, from time to time, supplying to the more refined agents designs they have found to yield high performance under their representations and loading conditions. Such designs must usually be automatically transformed from the representation of the "sending" agent to that of the "receiving" agent before that process can be performed. The net effect of this transfer of designs is to "draw the attention" of the more refined agents to the area of the design space represented by the transformed, less refined designs.

Since the less refined agents are searching in reduced design spaces and searching many times faster than the refined agents, they help the more refined agents avoid the huge waste of time represented by searching in unpromising regions of the design space, and help to concentrate their search where it will pay off with relatively few of the costly, full-refinement evaluations. At the same time, designs transferred in from different agents may possess different attractive attributes, and the search process in the more refined agent has the opportunity to seek designs incorporating the best features of each of the designs transferred in.

8.   EXAMPLE OF USE OF COLLABORATIVE INDEPENDENT AGENTS

To illustrate this process, another (single) automotive rail example will be used. The design space includes eleven sections similar to those illustrated

in Figure 3-5, with the gage thickness of the material (steel) as an additional
design variable (see Figure 3-7). There are a total of 67 design variables. The
complete, full-refinement design is described by the displacements of the spline-
controlling points in each of the sections, plus material gage. The performance
measures are as before – minimizing energy absorbed under several different
loading (test) conditions, with constraints on peak force, on manufacturability,
and on weight. A very simple three-agent model is shown in Figure 3-8. In that
scheme, the less refined agents explore only initial portions of the crash event,
reducing simulation time. They work in less refined design spaces, since "fine
tuning" will not be attempted in those two agents.

67 design variables: 66 control points and one gage thickness

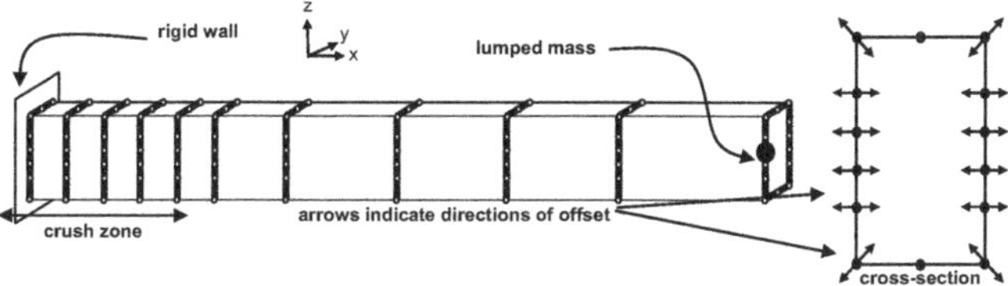

Figure 3-7.   Design variables in automotive rail problem for collaborative independent agents

This scheme, while easy to understand and offering some improvement in
search efficiency, is not adequate to allow rapid solution of this problem. Instead,
a larger set of agents is used (Figure 3-9), 19 in all, and the differences among
agents include: 1) whether or not they include stochastic variation of parameters
and loading conditions, 2) which loading cases are simulated, 3) resolution of
the design variables. The agents in the lower central compartment (numbers
10–17) represent the problem at its full refinement, and include consideration
of both load cases, also including stochastic variability of the design variables
and applied loads. The reason for using four agents in the center compartment is
to provide a variety of design "ideas" to the agents at the lowest level. They are
not allowed to exchange designs with each other in order to avoid any tendency

toward convergence on one region of the design space. The common influence exerted on all of them by having designs immigrating from the single, middle top-level agent is generally overcome in one of two ways: 1) when the run has progressed for a sufficient time, the top center agent's search has converged, and it may be shut down, and 2) similar designs repeatedly immigrating are "crowded" into the agent, tending to replace other designs most similar to them, so that the number of designs descending from a particular type of immigrating design is limited. The reason for using multiple (identical) agents in the bottom center compartment is in order to maintain the diversity of the most refined search, so that it does not quickly converge in the vicinity of the first promising designs found, but continues to combine "good design ideas" introduced from all of the other agents throughout the total time available for the search.

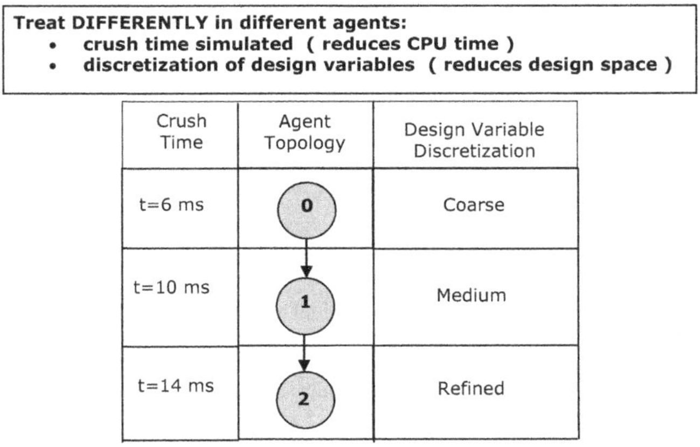

Treat DIFFERENTLY in different agents:
- **crush time simulated (reduces CPU time)**
- **discretization of design variables (reduces design space)**

Crush Time	Agent Topology	Design Variable Discretization
t=6 ms	0	Coarse
t=10 ms	1	Medium
t=14 ms	2	Refined

Figure 3-8.   Differences and communication among three example agents defined.

The 19-agent search shown in Figure 3-9 could be done on a single processor, but because each analysis required sizable amounts of time, the agents were distributed among 19 PCs on an Ethernet LAN.

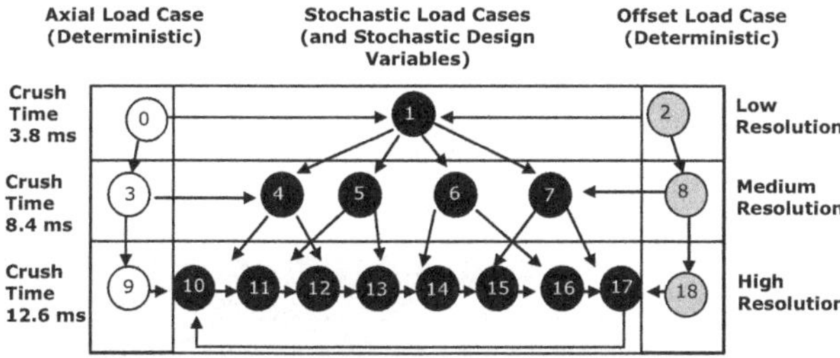

Figure 3-9.   Definition of the 19 agents used in design of a lower compartment rail.

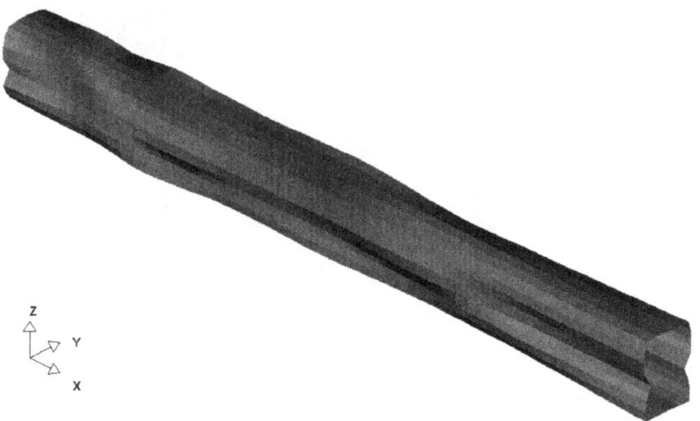

Figure 3-10.   Example of a rail resulting from the CIA design process (later rail design manufactured/tested is not shown because design was proprietary to customer).

Figure 3-10 shows the design of the rail that resulted from this search process, which required about one week on the 19 nodes. The rail produced was extremely novel, looking unlike any rail human designers had ever generated, and its calculated performance was decidedly better than those the designers had produced in their earlier manual attempts. According to the LS-DYNA analyses, peak force was reduced by 30%, energy absorption increased by 100%, weight reduced by 20%, and overall crash response would have resulted in the equivalent of a FIVE STAR rating. In subsequent (but proprietary) design work, similar approaches were used to generate a rail design that was manufactured in limited quantities and subjected to actual crash testing. As shown in Figure 3-11, the results of the crash test agreed very well with the LS-DYNA calculations, and the rail crushed as the FEA analysis had predicted.

Figure 3-11.　　Crash test of vehicle built with a rail developed using a CIA procedure similar to that described here. Crash behavior closely matched that predicted by the LS-DYNA FEA model.

9.　　SUMMARY

Two methods have been described that, together, offer an approach to optimizing the design of many systems and/or components that are generally not seen as practical to submit to computational optimization. The first, COMPOSE, is useful for designing components or subsystems of larger systems which are tightly coupled through shared boundary conditions that are strongly affected by changes in the component design. Use of COMPOSE may allow reducing the number of whole-system evaluations necessary for component design from hundreds or thousands to fewer than ten, while still exploring the component design space extensively. The second, collaborative independent agents, allows addressing problems that have both large design spaces and time-intensive analyses, rendering them intractable to traditional methods. Through careful design of a set of loosely coupled optimization agents, computing time to find good solutions to such problems may be dramatically reduced. The savings results from the continuing transfer of results from rapid, low-refinement, less accurate search to agents that search at the full level of refinement and accuracy demanded of a solution of the problem, essentially providing them guidance as to what portions of the design space are likely to be worth searching in greater detail.

References

Averill RC, Eby D and Goodman ED (2001) How well can it take a hit? – An advanced automated optimization technique can help designers develop crashworthy structures. *ASME Mechanical Engineering Design*, March, pp. 26–28.

Averill RC, Punch WF, Goodman ED, Lin S-C, Yip YC, Ding Y (1995) Genetic algorithm-based design of energy-absorbing laminated composite beams. *Proc. ASME Design Eng. Tech. Conf.*, Vol. 1, Boston, pp. 89–96.

Chapman CD, K. Saitou K and Jakiela MJ (1993) Genetic algorithms as an approach to configuration and topology design. Proc. *1993 ASME Design Automation Conference*, Albuquerque, New Mex., Sept.

Daratech Research Staff (2002) Process Integration, Design Optimization Emerging as New Solution Category.

Eby D, Averill RC, Goodman, ED and Punch W (1999) The optimization of flywheels using an injection island genetic algorithm. *Evolutionary Design by Computers*, P. Bentley, ed., Morgan Kaufmann, San Francisco, pp. 167–190.

Eby D, Averill RC, Punch WF, Mathews O, and Goodman ED (1997) An island injection GA for flywheel design optimization. *Proc. EUFIT '97*, Aachen, Germany, pp. 687–691.

Eby D, Averill RC, Punch WF and Goodman E (1998) Evaluation of injection island GA performance on flywheel design optimization. *Adaptive Computing in Design and Manufacture*, I. C. Parmee, ed., Springer, Berlin, pp. 121–136.

Haftka RT (1984) An Improved Computational Approach for Multilevel Optimum Design. *J. Structural Mechanics*, Vol. 12, pp. 245–261.

Leung M and Nevill GE, Jr. (1994) Genetic algorithms for preliminary 2-D structural design. *Proc. 35th AIAA/ASME/AHS SDM Conf.*, Hilton Head, SC, April 18–20.

LeRiche R and Haftka RT (1993) Optimization of laminate stacking sequence for buckling load maximization by genetic algorithm. *AIAA Journal*, 31, pp. 951–956.

Lin S-C, Punch WF and Goodman ED (1994) Coarse-grain parallel genetic algorithms: categorization and analysis. *Proc. IEEE Symposium on Parallel and Distributed Processing*, pp.27–36.

Malott B, Averill RC, Goodman ED, Ding Y, and Punch WF (1996) Use of genetic algorithms for optimal design of laminated composite sandwich panels with bending-twisting coupling. *AIAA/ASME/ASCE/AHS/ASC 37th Structures, Structural Dynamics and Materials Conf.*, Salt Lake City, Utah.

Nagendra S, Haftka RT, and Gurdal Z (1992) Stacking sequence optimization of simply supported laminates with stability and strain constraints. *AIAA Journal*, 30, pp. 2132–2137.

Nagendra S, Haftka RT, and Gurdal Z (1993) Design of blade stiffened composite panels by a genetic algorithm approach. *Proc. 34th AIAA/ASME/AHS SDM Conference*, La Jolla, CA, April 19–22, pp. 2418–2436.

Papadrakakis M and Tsompanakis Y (1999) Domain decomposition methods for parallel solution of shape sensitivity analysis problems *Int. J. Num. Meth. Eng.*, Vol. 44, pp. 281–303.

Red Cedar Technology, proprietary internal report, 2001.

Sandgren E, Jensen E, and Welton JW (1990) Topological design of structural components using genetic optimization methods. *Sensitivity Analysis and Optimization with Numerical Methods*, Saigal S and Mukherjee S, eds, AMD-Vol. 115, ASME, pp. 31–43.

Sobieszczanski-Sobieski J, Altus TD, Phillips M and Sandusky R (2003) Bilevel integrated system synthesis for concurrent and distributed processing. *AIAA Journal*, Vol. 41, pp. 1996–2003.

Sobieszczanski-Sobieski J, James BB and Dovi RR (1985) Structural optimization by multilevel decomposition. *AIAA Journal*, Vol. 23, No. 11, pp. 1775–1782.

Sobieszczanski-Sobieski J, James BB and Riley MF (1987) Structural sizing by generalized, multilevel optimization. *AIAA Journal*, Vol. 25, No. 1, pp. 139–145.

Soto CA and Diaz AR (1993) Optimum layout and shape of plate structures using homogenization. *Topology Design of Structures*, Bendsoe MP and Mota Soares CA, eds., pp. 407–420.

Suzuki K and Kikuchi N (1990) Shape and topology optimization by a homogenization method. *Sensitivity Analysis and Optimization with Numerical Methods*, AMD-Vol. 115, ASME, pp. 15–30.

Suzuki K and Kikuchi N (1991) A homogenization method for shape and topology optimization. *Comp. Meth. A*

Chapter 4

REPRESENTING THE CHANGE -
FREE FORM DEFORMATION FOR
EVOLUTIONARY DESIGN OPTIMIZATION

Stefan Menzel[1] and Bernhard Sendhoff[1]

[1]*Honda Research Institute Europe GmbH, Carl-Legien-Str. 30,
D-63073 Offenbach/Main, Germany*

Abstract   The representation of a design influences the success of any kind of optimization significantly. The perfect trade-off between the number of parameters which define the search space and the achievable design flexibility is very crucial since it influences the convergence speed of the chosen optimization algorithm as well as the possibility to find the design which provides the best performance. Classical methods mostly define the design directly, e.g. via spline surfaces or by representations which are specialized to one design task. In the present chapter, the so-called *deformation methods* are focused which follow a different approach. Instead of describing the shape directly, deformation terms are used to morph an initial design into new ones. This decouples a complex design from an expensive shape description while relying purely on mapping terms which are responsible for the geometry transformations. Thus, the designer is encouraged to determine the optimal relation between parameter set and design flexibility according to the given task. With respect to the optimization, these mapping terms are considered as parameters. In this chapter, the combination of two state of the art deformation algorithms with evolutionary optimization is focused. After an introduction of these techniques, a framework for an autonomous design optimization is sketched in more detail. By means of two optimizations, which feature a stator blade of a jet turbine the workability is shown and the advantages of such representations are highlighted.

Keywords: design optimization, direct manipulation, evolutionary algorithms, evolutionary optimization, free form deformation, representation, turbine blade

S. Menzel and B. Sendhoff: *Representing the Change - Free Form Deformation for Evolutionary Design Optimization*,
Studies in Computational Intelligence (SCI) **88**, 63–86 (2008)
www.springerlink.com   © Springer-Verlag Berlin Heidelberg 2008

1.   INTRODUCTION

Evolutionary algorithms have been successfully applied to a variety of design optimization problems (Sonoda et al., 2004; Oyama et al., 2000; Foli et al., 2006; Ong et al., 2006; Kanazaki et al., 2002). In most cases, the design is represented either by a specialized representation based on standard engineering practice, e.g. circles and connecting curves for two-dimensional blades or an aircraft wing design (Oyama et al., 2000), or by splines, NURBS (Hasenjäger et al., 2005; Lépine et al., 2001) and D-NURBS (Terzopoulos et al., 1994). Other representations like solid modelling (Requicha, 1980) or solutions to partial differential equations (Ugail et al., 2003) have been used less frequently, especially for practical applications.

While standard engineering representations are often very compact, they are almost always incomplete in the sense that not all possible shapes can be represented. Although NURBS are only complete if the representation is adaptive (Olhofer et al., 2001), their versatility is usually much higher. At the same time, the representation of complex shapes with many edges (e.g. a whole turbine) requires a NURBS surface with a very large number of control points. However, the number of control points defines the dimension of the search space. If the dimension is much larger than 200 and no subspaces can be easily identified to allow a sequential search, the shape design optimization process is prohibitively time consuming and the convergence may not be achieved.

Free form deformation (FFD) techniques, which we will introduce in the next section, are fundamentally different from other representations in that they do not represent the actual shape but changes to a baseline shape. The baseline shape can be arbitrarily complex. Of course the changes that can be represented might be limited by the complexity of the representation. In this chapter, we will introduce free-form deformation and direct manipulation of free-from deformation in the context of evolutionary shape design. We will apply both methods to the design of a stator blade for a real gas turbine engine. Although the strength of FFD is particularly evident for very complex shapes, turbine blades are a reasonable compromise, because their complexity is high enough to require a NURBS representation with just below 100 control points (Hasenjäger et al., 2005) while it is sufficiently simple to analyze the evolutionary process.

2. AERODYNAMIC OPTIMIZATION USING DEFORMATION TECHNIQUES

In order to realize a fully automated aerodynamic design optimization, the so-called *deformation techniques* are becoming more widely used as an efficient object representation recently. Introduced in the late 1980's in the field of computer graphics (Sederberg et al., 1986; Coquillart, 1990), these methods hold several advantages for representing geometries, especially if these geometries possess a high degree of complexity. Usually the number of parameters to represent such complex geometries, e.g. only by splines or spline surfaces, is too large to be feasible. When applying free form deformation (FFD), the current state-of-the-art deformation algorithm, the object is embedded within a lattice of control points, which defines the degrees of freedom for the deformation. The parameter set and consequently the difficulty of the optimization problem can be tuned by the number and choice of control points or control point groups. When defining the parameter set of a FFD system we have to find the optimal trade-off between search space dimension and design flexibility, i.e. freedom of variation. However, compared to spline based representations, in the FFD framework, the complexity of design variations and not of the initial design is the limiting factor.

In case of problems that need finite element or finite volume methods for design evaluations like in computational fluid dynamics, FFD has another advantage (Perry et al., 2000; Menzel et al., 2005). The fidelity of computational fluid dynamics (CFD) simulations depends to a large degree on the quality of the mesh or grid that is used for the simulation. For complex shapes and structures, mesh generation is a very time consuming process (several days), which more often than not requires manual fine-tuning or resolution of meshing problems. In particular, in the context of population based search methods like evolutionary algorithms, manual mesh generation is not feasible. In the FFD framework, the mesh is deformed just like the design is deformed. Therefore, the mathematical deformation procedure is applied to the design and to the mesh simultaneously. This has the great advantage that mesh generation just has to be done once at the beginning of the optimization for the baseline design. During optimization, the mesh is always adapted to the changing design. Of course, the quality of the deformed mesh has to be controlled during the evolutionary search process. However, for complex shape design, the whole optimization process takes days and weeks and the sporadic analysis of the current mesh quality in parallel to the search process is feasible.

In Section 3, two kinds of deformation methods are briefly introduced: the standard free form deformation (FFD) and the direct manipulation of free form deformations (DMFFD) as one of its extensions. The latter method makes use of object points as direct handles on the geometry to take into account the sensitivity of the FFD method to the initial placement of the control points. In Section 4, both methods are used as representations for an evolutionary design optimization of an aerodynamic shape case study. On the one hand FFD is applied to optimise a three-dimensional blade geometry to show the basic behavior of FFD in optimizations while dealing with complex structures. On the other hand both techniques, FFD and DMFFD, are used as representations in a two dimensional blade optimization to illustrate the differences and advantages of DMFFD over FFD.

3.   REPRESENTING DESIGNS WITH DEFORMATION METHODS

In this section, the methods of free form deformation (FFD) and direct manipulation of free form deformations (DMFFD) are briefly introduced.

3.1   Free Form Deformation (FFD)

The basic idea behind free form deformation is depicted in Figure 4-1 a). The sphere represents the object which is the target of the optimization. It is embedded in a lattice of control points (CP). Firstly, the coordinates of the object have to be mapped to the coordinates in the spline parameter space. If the object is a surface point cloud of the design or a mesh which originates from an aerodynamic computer simulation (as in our example in Section 4), each grid point has to be converted into spline parameter space to allow the deformations. After this process of 'freezing', the object can be modified by moving a control point to a new position. The new control point positions are the inputs for the spline equations and the updated geometry is calculated. Since everything within the control volume is deformed, a grid from computational fluid dynamics that is attached to the shape is also adapted. Hence, the deformation affects not only the shape of the design but also the grid points of the computational mesh, which is needed for the CFD evaluations of the proposed designs. The new shape and the corresponding CFD mesh are generated at the same time without the need for an automated or manual re-meshing procedure. This feature significantly reduces the computational costs and allows a high degree of automation. Thus, by applying FFD, the grid point coordinates are changed but the grid structure is kept.

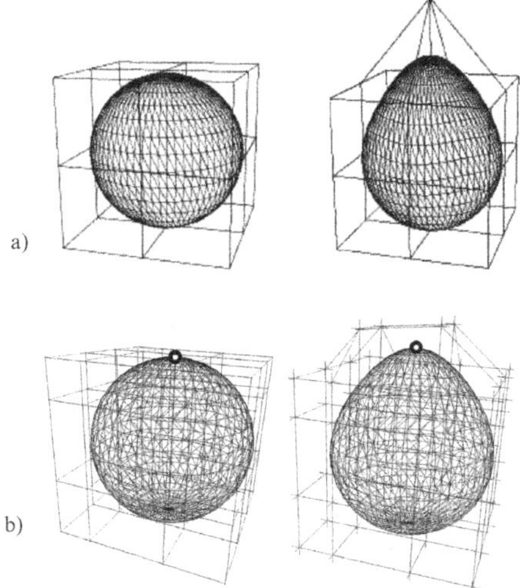

Figure 4-1. a) Free Form Deformation (Perry et al., 2000). The design is embedded within a lattice of control points. The modification of control points affects the shape as well as everything else inside the control volume. – b) Direct Manipulation of Free Form Deformations. The object point is chosen directly on the surface and the required movements of the control points to realize the target movement of the object point are calculated e.g. by the least squares method. The dotted control volume is invisible to the designer as s/he works directly on the object points; the control volume can be chosen arbitrarily.

As we already mentioned, the number and distribution of control points have to be chosen carefully. As one can imagine, an inappropriate set-up of the FFD control volume increases the necessary size of the parameter set and, therefore, the dimensionality of the search space. One of the reasons is that the impact of a control point on an object decreases when the distance from the object increases. Even a small object variation requires a large modification of the control point if the initial distance between object and control point is large which also violates the strong causality condition that is important in particular for Evolution Strategies (Sendhoff et al., 1997). This in turn often modifies other areas of the design space which has to be compensated for by the movement of other control points. Hence, often correlated mutations of control points are necessary for a local change of the object geometry. To reduce the influence of the initial positions of the control points, DMFFD is considered as a representation for evolutionary optimization. DMFFD allows to determine variations directly on the shape. Therefore, local deformations of the object depend only on the so called object points.

3.2 Direct Manipulation of Free Form Deformations (DMFFD)

Direct manipulation of free form deformations as an extension to the standard FFD has been introduced in (Hsu et al., 1992). Instead of moving control points (CP), whose influence on the shape is not always intuitive, the designer is encouraged to modify the shape directly by specifying so called object points (OP).

Although the initial setup of the control volume is similar to FFD, the control volume becomes invisible to the user and necessary correlated modifications are calculated analytically. In a first step, a lattice of control points has to be constructed and the coordinates of the object and the CFD mesh have to be frozen. But the control volume can be arbitrary, i.e., the number and positions of control points do not need to have any logical relationship to the embedded object, besides the fact that the number of control points determines the degree of freedom of the modifications. In the next step, the designer specifies object points, which define handles of the represented object that can be repositioned.

The shape is modified by directly changing the positions of these object points. The control points are determined analytically so that the shape variations (induced by the object point variations) are realized by the deformations associated with the new control point positions. In other words, the control points are calculated in such a way that the object points meet the given new positions under the constraint of minimal movement of the control points in a least square sense. Of course the object variations must be realizable by the deformations from the calculated new control point positions, i.e., if the number of control points is too small, some variations given by new object point positions might not be representable by a deformation.

In Figure 4-1 b) an object point has been specified at the top of the sphere. The designer is able to move this object point upwards without any knowledge of the "underlying" control volume which can be initialized arbitrarily. The direct manipulation algorithm calculates the corresponding positions of the control points to mimic the targeted object point movement. The solution is shown in Figure 4-1 b).

Direct manipulation of free form deformation has several advantages when combined with evolutionary optimization as compared to standard FFD. First, the construction of the control volume and the number and distribution of control points are not as important as in standard FFD. Furthermore, the number of optimization parameters equals to the number of object points. For an illustration of both methods in the context of an evolutionary design optimization, a stator blade of a jet turbine is considered as a test scenario. The set-up and the results are discussed in Section 4.

4. FFD AND DMFFD IN EVOLUTIONARY DESIGN OPTIMIZATION

In this section, two applications of FFD methods for the evolutionary optimization of aerodynamic structures are described. The aerodynamic problems are well suited to demonstrate both advantages of FFD. They are sufficiently complex and they require CFD calculations for the evaluation.

Evolutionary algorithms belong to the group of stochastic optimization algorithms. They mimic the principles of Neo-Darwinian evolution, see e.g. (Fogel, 1995; Rechenberg, 1994; Schwefel, 1995) by applying operators for reproduction, mutation and/or recombination and selection. Prominent examples of EAs are Evolution Strategies (ES), Genetic Algorithms (GA) or Genetic Programming (GP). Among the advantages of evolutionary algorithms are robustness against noisy or discontinuous quality functions, the ability to escape from local optima and to enable global search. In the course of optimization, a population of possible solutions (e.g. a vector of continuous parameters, the objective variables) keeps adapted to solve a given problem over several generations. The adaptation occurs by variation of solutions contained in a population and by selection of the best solutions for the next generation.

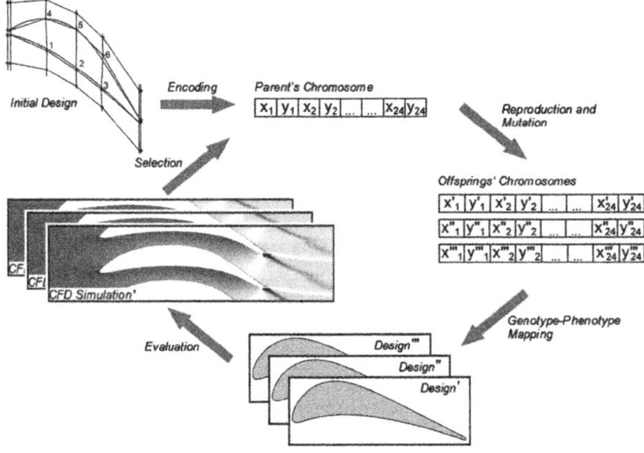

Figure 4-2. The generational cycle in evolutionary design optimization.

Schematically, the evolution cycle is depicted in Figure 4-2, which has already included the turbine blade as the optimization target. In this paper, a special variant of Evolution Strategies, the Covariance Matrix Adaptation (CMA), is applied which has the advantage of a high convergence rate for real-valued problems compared to other evolutionary algorithms. This is particularly important for very time consuming evaluations like CFD simulations. The successful application of this type of algorithm has been shown previously e.g. for a two-dimensional turbine blade optimization (Sonoda et al., 2004; Olhofer et al., 2001). A detailed description of the CMA algorithm is provided in (Hansen et al., 2001).

4.1   The Stator Turbine Blade as an Aerodynamic Test Scenario

The subject of optimization in this study is a turbine stator blade that is part of a gas turbine for a small business jet. An illustration of the turbine is shown in Figure 4-3. Around the hub of the turbine, eight blades are equally distributed. Because of the low number of stator blades, this design is referred to as a ultra-low-aspect-ratio stator and is less common for the design of gas turbine engines. For more detailed information on the turbine architecture and on the results of a spline-based optimization, the interested reader is referred to (Hasenjäger et al., 2005). Here we will use the design problem of a three dimensional stator blade to highlight the applicability of the standard FFD for complex structures, and a two-dimensional optimization to compare DMFFD with FFD.

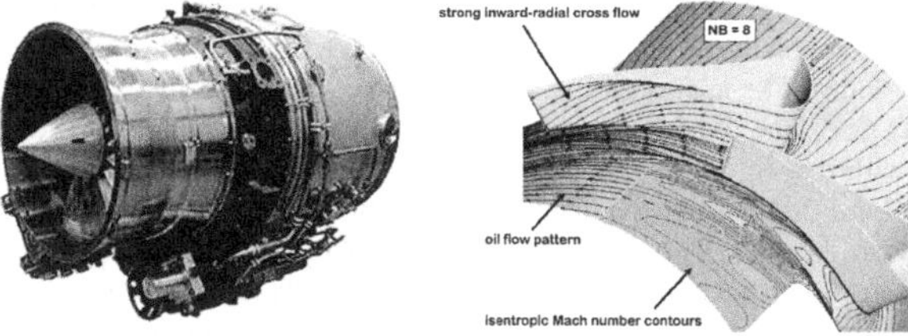

Figure 4-3.   Gas Turbine and its fluid dynamics in one blade section (Hasenjäger et al., 2005).

4.2   Evolutionary Optimization of a 3D Structure using FFD

The first step when it comes to numerical design optimization is to extract the characteristic optimization parameters from the representation of the given geometry. In terms of applying free form deformation, as mentioned above, a lattice of control points has to be constructed which encloses the target geometry. Because of the rotational symmetry in the present problem, only one of the eight turbine blade sections needs to be extracted for further evaluations. In the present problem, the fitness value is calculated via CFD simulation and the region between two blades has to be meshed for solving the numerical equations accordingly. This mesh, which includes the blade's suction and pressure side as boundary layers, was embedded by the lattice of control points to allow the simultaneous deformation of stator blade and computational grid. Finally, twelve control points have been taken from the lattice as optimization parameters. To simplify the calculations and because of the bending of the turbine blade the global x, y and z coordinates of the design and of the knots of the CFD grid have been transferred to a local cylinder coordinate system x', y' and z'. The lattice is fully three dimensional and a sample cross-section is depicted in Figure 4-4. As explained above, the CFD mesh plays an important role in the blade optimization and, therefore, all grid knots that can be found in the CFD mesh between two neighboring blade surfaces have to be fully embedded in the control volume. As a consequence, the deformations are applied to the turbine blade surfaces and simultaneously to the CFD mesh so that a re-meshing process can be omitted.

The blade shape is depicted in Figure 4-4. The continuous upper line is the pressure side and the continuous lower line the suction side of two neighboring turbine blades. The grey region marks the area of the knots and volume cells of the CFD grid. The two blade contours are depicted to show the position of the blades with respect to the control volume. As already mentioned, it should be kept in mind that not the shape of one whole blade is embedded in the control volume but the passage between two blades where the CFD grid is defined. In local x'-direction seven control points have been placed. In local y'-direction the rotational symmetry strongly influences the number and positioning of the control points.

Although all of these control points are important for freezing and deforming the geometry and the CFD mesh, only 12 points have been optimised. Six of these points are shown in Figure 4-2 and Figure 4-4 for the hub section and another six points have been chosen analogously for the casing section. In total 24 parameters (x and y coordinates of the 12 points) have been considered in the evolutionary optimization and were encoded in the parent's chromosome.

To maximize the influence of these control points on the blade geometry they have been positioned as close as possible to the boundary layers of the blade so that the mutation of the control points has a high impact on the design. In this first test scenario due to the small population size the number of optimization parameters has been kept as low as possible. As a consequence, the movements of the 12 control points result in rather global design changes of the blade, which can also be observed from the resulting shapes depicted in Figure 4-6. The present optimization was motivated by the need to correlate these global design changes with the performance differences. A higher degree of locality of the changes can easily be realized by refining the control point lattice by increasing the number of control points, e.g. the optimization with the spline representation discussed in (Hasenjäger et al., 2005) required 88 parameters.

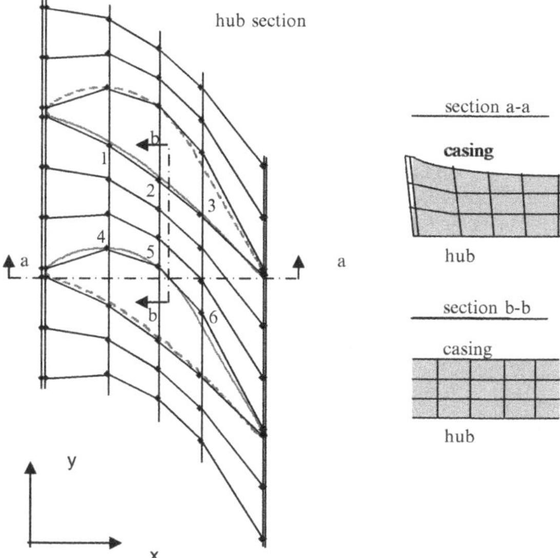

Figure 4-4. Embedding the turbine blade and the CFD mesh in an FFD lattice.

Based on these control point settings the initial CFD grid and the blade geometry have been frozen, i.e. the coordinates of the grid knots have been calculated in spline parameter space. Afterwards the 24 parameters were encoded in the initial parent's chromosome and the optimization was started. The first parameter sets were generated and extracted to calculate the new positions of the control points. Based on these updated control point positions the free form deformation of the CFD grid and the blade geometry were performed and the CFD simulation started.

After the calculation has finished the result of the fitness function is determined by a weighted sum of two flow features and three geometric properties. As the main optimization criterion the minimization of the pressure loss has been chosen. To keep the blade geometry within feasible constraints, four additional values have been extracted from the CFD calculation and blade geometry, respectively. Often an optimum of the fitness landscape is very close to the constraints, hence the boundary conditions have to be checked carefully. To avoid hard constraints, which would directly exclude illegal designs, weights have been introduced so that it was possible to determine a performance index for all evaluations. High penalty terms ($1e^{20}$) have been assigned to these weights w_2 to w_5 which outweigh the contribution of the objective t_1 by far. In case of a violation of constraints, the optimization is quickly driven back to feasible design regions. Before the optimization, the target ranges for the outflow angle, the maximum solidity, the minimum blade thickness and the trailing edge thickness have been defined observing constraints set by other turbine parts, by used materials and by the manufacturing process. Whereas the solidity t_3 of the turbine is a measure for the blade spacing, the blade thickness t_4 and trailing edge thickness t_5 are calculated for a single blade. The calculation is done as follows:

$$f = t_1 + \sum_{i=2}^{5} w_i t_i^2 \rightarrow \min \tag{1}$$

with:

t_1 pressure loss t_2 difference to target outflow angle
t_3 difference to target solidity
t_4 difference to target minimum blade thickness
t_5 difference to target minimum trailing edge thickness
w_i weights for the different input data $t_2, \dots t_5$

The course of the fitness is depicted in Figure 4-5. Note that the fitness is to be minimized. A total number of 134 generations has been calculated resulting in an overall optimization time of approximately six weeks on a computer cluster. The runtime is closely related to the performance of the CFD solver which depends on the calculation models and computational grid. The overall grid size of one simulation was $175 \times 52 \times 64 = 582400$ cells and the time for the calculation of one blade took about five to six hours on a PIII Xeon, 2.0 Ghz node, depending on the convergence behaviour. As flow solver, the parallelized 3D Navier-Stokes flow solver HSTAR3D (Arima et al., 1999) has been used which is perfectly adjusted to the present problem. The solver is parallelized for four CPUs resulting in a total usage of 8 individuals $\times 4$ CPUs $= 32$ CPUs at the same time. The node communication was realized via the Parallel Virtual Machine (PVM) framework in a master/client configuration (Geist at al., 1995).

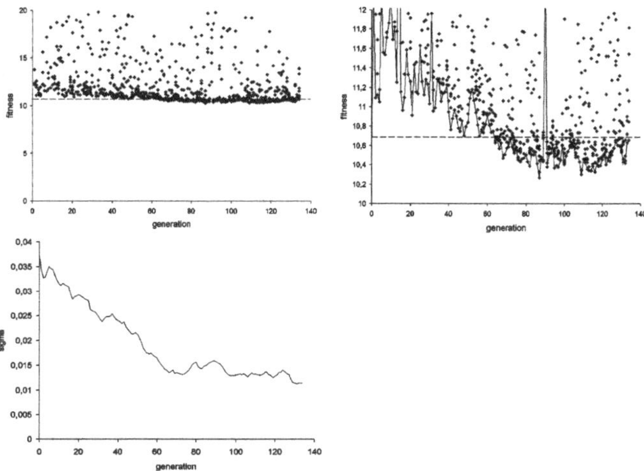

Figure 4-5.   Courses of fitness and global step-size during optimization.

In the first ten generations a (1,6)-strategy has been used but was extended to a (1,8)-strategy starting from generation 11 because of the high variance of the fitness values. Generally, a population size above 10 is recommended but could not be realized due to restrictions of the available computer power. The fitness value of the initial blade is about 10.69 and is marked by the dashed line in the fitness graphs.

It can be seen that in the beginning the fitness value of the best design in each generation increases (no elitism) and reaches the initial level again at generation 60. After this point the best individual always performs better than the initial one and stays within a range of 10,27 to 10,60. The best value of 10,27 is reached in generation 87, which corresponds to a performance gain of 4 %.

The course of the optimization can also be analyzed by observing the development of the global step-size. Right from the beginning, the step-size decreases and reaches a plateau after approximately 60 generations. At the beginning, large mutations were generated leading to an increase of the fitness value. This posed a serious problem due to the small population size of only 6 offspring. Therefore, the population size was increased to 8 individuals starting from generation 11. The initial blade and the shape of the best design from generation 87 are depicted in Figure 4-6 to visualize the changes which occurred.

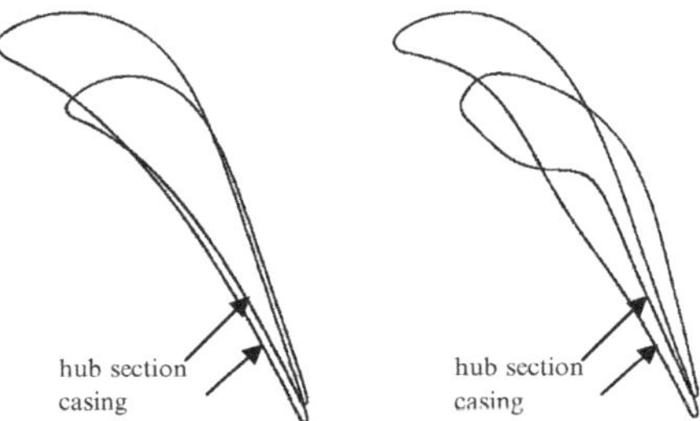

Figure 4-6.   Initial and optimized shape of the turbine blade (hub and casing section).

4.3 The Impact of Object Points: A Comparison of FFD and DMFFD as Representations in Evolutionary Design Optimization

As already explained in Section 3, the standard FFD optimization strongly depends on an appropriate set-up of the control volume which relies on the existing know-how of the designer. DMFFD can reduce this influence by introducing object points that can be placed directly on the shape.

In order to compare FFD with DMFFD, we have carried out four optimization runs of a two-dimensional turbine blade (see Figure 4-2). The first optimization uses the standard free form deformation representation and the remaining three the direct manipulation technique. The two dimensional scenario has been chosen because of the large amount of computational resources that are needed for the CFD simulations, especially for the three dimensional flow solver. In all four optimizations, the population size has been set to 32 individuals and an approximation model has been used. In a pre-evaluation step, all 32 individuals have been evaluated with a neural network and only the 16 most promising ones have been simulated with the CFD solver to determine the individual fitness. The "true" fitness values have also been used to re-train the neural network online. From the 16 CFD results the best individual has been selected as the parent for the next generation, similar to the standard notation of evolutionary strategies we call this a (1,32(16)) strategy.

Table 4-1. Type and number of parameters.

Run	Type	Number of parameters	Number of control points
1	control points	10	10
2	object points	5	10
3	object points	13	10
4	object points	13	36

The details for each run are summarized in Table 4-1. The number of parameters equals to the dimension of the search space. Their distribution on the design is depicted in Figure 4-7. The number of control points refers to the control point coordinates which can be modified in the FFD control volume. This is different from the total number of control point coordinates because points at the upper, lower and left border have to be constant due to CFD mesh consistency. Additionally, control points on the right edge of the control volume can be modified only in y-direction in order to fix the x-length of the design. For run 1 to run 3 the same FFD control mesh is used which is shown in Figure 4-7 in the upper left part for run 1.

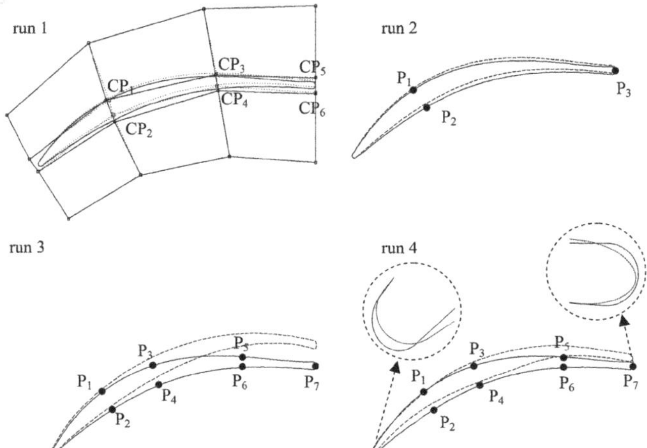

Figure 4-7. Number and distribution of optimization parameters. run1: 10 parameters (P_1–P_4: x, y; P_5, P_6: y); run2: 5 parameters (P_1, P_2: x, y; P_3: y); run3: 13 parameters (P_1–P_6: x, y; P_7: y); run4: 13 parameters (P_1–P_6: x, y; P_7: y). The continuous curve marks the initial designs, the dashed curves the optimized ones. The control volume is only drawn for run 1. The control volumes for run 2 and run 3 are the same as for run 1. Run 4 has been modified in such a way that two rows and columns of control points have been inserted corresponding to a simple knot insertion algorithms as explained in (Piegl et al., 1997).

The general workflow of the design optimization is similar to the optimization described in Section 4.2. A control volume consisting of 4×4 control points has been set up in which the turbine blade is embedded (see Figure 4-7). For easier visualization, the CFD mesh is not plotted. However, we should keep in mind that during the deformation step the blade geometry *as well as* the CFD mesh are modified which allows the omission of the costly re-meshing process.

The control points CP_1–CP_4 can be freely moved in the x-y plane during the optimization, while CP_5 and CP_6 are only allowed to move in the vertical direction as stated above. After the encoding of these parameters (x and y coordinates of points CP_1 to CP_4 and the y-coordinate of CP_5 and CP_6) in the chromosome of the parent individual, the control point positions are optimized. This includes the mutation of the control points, the deformation of the CFD grid based on the free form deformation algorithm and the execution of the CFD flow solver. As described in the previous section, the ES-CMA is used together with a neural network meta-model.

For run 4, the modifications at the leading and trailing edge are shown in a higher resolution to illustrate the occurring deformations. Initial circular or ellipsoid arcs are not kept after deformation because they turn out to be inferior to other leading and trailing edge geometries.

In runs 2, 3 and 4, the direct manipulation of the free form deformations is applied to modify the control points directly, i.e.:

1. The chromosomes contain object point positions (P_i) instead of control point positions (CP_i) as parameter sets.

2. The control points are calculated based on the encoded object points with the method for direct manipulation. Here the object points given in Figure 4-7 are used in the three runs. According to Section 3, the procedure of how to calculate the control point positions which are required for deforming the design and the grid is sketched as follows. After the object point positions have been mutated in each generation, the positions of the control points are updated. The new positions of the control points are calculated in such a way that the modifications of the object points are realized as best as possible, i.e. in the present scenario in a least squares sense as described in (Hsu et al., 1992). After an update of the control volume, the design and the CFD grid is deformed and prepared for performance evaluation.

The fitness progressions of all four optimization runs are summarized in Figure 4-8.

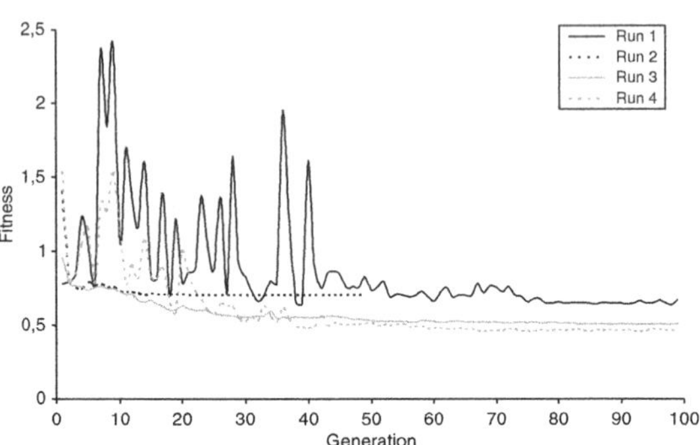

Figure 4-8.   The progress of the fitness of the runs 1–4.

One major drawback of the direct manipulation method is that the calculation of the control points has to be carried out based on the desired object point positions without being able to incorporate constraints imposed by the CFD grid. In particularly, negative volumes can emerge which can be described as loops in the design space.

This can usually be avoided by keeping the order of control points during the deformation step. However, when using direct manipulation, a desired object point position can sometimes only be realized by a large degree of control point modifications including destroying the order of control points. Methods for repairing and improving the structure of control points are therefore topic of our current research (Bihrer, 2006). To guarantee valid CFD meshes in the present optimization the order of control points is checked after every mutation step. If the order of the control points is changed the mutation is repeated until a valid individual is generated. Figure 4-8 summarizes the results of the optimizations. Base run 1 that uses the standard FFD representation resulted in a converged fitness of 0.62 which means a 37% gain compared to the fitness of the initial turbine blade of 0.98.

According to Figure 4-7, three object points have been chosen for run 2. It resulted in a fitness of 0.7 but it needed less than half the number of generations and the optimization run is very stable. This is due to the reduced number of parameters which is only 5 (2 object points movable in x- and y-, one object point movable in y-direction). However, it also shows that the flexibility of the design is limited by the choice of object points. This demonstrates that an optimization using direct manipulation is limited by two factors. On the one hand, a low number of object points restricts the flexibility of the design because these are the parameters which are optimized. On the other hand, the number of control points limits the degree of realizable shape variations because the control points *actually* induce the targeted object point modifications through the defined deformations. If the number of control points is too low, the targeted object point movements cannot be achieved.

In run 3, the number of object points (OP) has been increased to 7, i.e. 13 optimization parameters (6 OP movable in x- and y-, one OP movable in y-direction) to improve the flexibility of the design. The fitness decreased to 0.5. This is an improvement compared to the optimization run 1. This improvement is particularly interesting because the optimization is based on the same control point grid as that in run 1. Even if the number of parameters for the optimization is larger than in run 3, the parameters for the deformations are identical because they are limited by the control point grid. Therefore, the actual design freedom is the same in both runs. Since the number and distribution of control points did not change between both runs the optimization of run 1 must have converged to a local optimum. The structure of the search space seems to be changed by the direct manipulation in a way that the local maximum is avoided in this optimization run. Of course we must be careful to draw conclusions from just one optimization run. Nevertheless, we can state that the different representation of the same degree of variability will lead, in general, to a different search space behavior.

As a consequence, for this optimization it can be seen that the usage of object points has been more successful. The fitness decreased faster and also at an earlier generation which is particularly important when dealing with time consuming evaluation functions like CFD simulations.

To analyze whether the performance could be even more increased by allowing more flexibility in the possible deformations, two rows and columns of control points have been inserted into the control volume, resulting in 36 control points in run 4 while the number of object points was kept at 7. The fitness improvement due to the control point insertion is only slightly increased to 0.45. This is also a promising observation because the number of optimization parameters is still 13 and the course of the fitness is quite similar to the one of run 3. Hence, the increase of flexibility by control point insertion did not affect the convergence behavior.

In summary, we have shown that the usage of the direct manipulation with free form deformation method has been advantageous in many ways in this optimization. If only 3 object points are chosen, like in run 2, the convergence speed improved drastically and resulted still in a good performance, compared to the optimization of the control points in run 1. This can be explained by the lower number of parameters in the optimization. If the number of object points is increased, like in run 3, and at the same time keeping the control points fixed, the fitness can be further improved although the possible transformations are kept constant in all three experiments. Here obviously the re-structuring of the search space by the introduction of the direct manipulation methods is beneficial.

Even an increase of control points in the control volume, as it has been done in run 4, did not slow down the optimization. This is a very promising result since the influence of the number of control points did not affect the convergence speed but the number of object points did. As a consequence one could argue for choosing a high number of control points in the optimization to achieve a high flexibility of the transformation and less constraints for the modification due to restrictions in the transformation. This definitely decreases the effect of the control point position and reduces the necessary prior knowledge about the optimization problem while setting up the control volume.

5. SUMMARY AND CONCLUSIONS

In this chapter, the features and advantages of the application of deformation techniques as a representation in evolutionary design optimization have been presented. Even complex designs like automobile parts can be encoded by free form deformation techniques. The limiting factor is not the complexity of the baseline design but the allowed variations. The fewer the control points, the more global the induced changes. If the control points are positioned based on expert knowledge, even meaningful large scale changes could be represented. At the same time, if a large number of control points is available, local variations can also be realized just as with standard spline representations. For the future, one could envision hierarchical FFD representations which can incorporate (and rely on) expert knowledge to a different degree.

The second decisive advantage of FFD representations for evolutionary design optimization is that the computational mesh for the CFD calculations is deformed together with the design shape. Therefore, a costly re-meshing procedure can be avoided. Indeed for some very complex geometries, re-meshing during optimization is not possible and in turn optimization is only possible with FFD representations. Of course it is still necessary to check that the mesh remains to be feasible after a certain number of deformation e.g. every tenth generation.

Besides applying FFD representations to the evolutionary design optimization of a three-dimensional stator blade of a gas turbine engine, to demonstrate the feasibility in a reasonably complex test scenario, we also introduced direct manipulation FFD as an extension. In DMFFD, design changes are only indirectly encoded in the chromosome. The evolutionary optimization acts directly on object points, however, this only leads to "desired" design variations. These "desired" changes are then realized as close as possible by the underlying FFD algorithm based on a certain number of control points using e.g. a least squares algorithm. Three scenarios are possible. (1) The desired degree of freedom is larger than the realizable degree of freedom – thus evolutionary induced changes might not be realized; (2) the desired degree of freedom roughly equals the realizable degree of freedom – thus most changes can be realized one to one; (3) the desired degree of freedom is smaller than the realizable degree of freedom – thus desired changes can be represented in different ways and therefore, different path' through the search space are available. All three relations are interesting in their own right and deserve a more detailed analysis.

A more practical problem of DMFFD is that constraints for mesh deformation are more difficult to incorporate in the search process. Additional methods for securing mesh consistency must be researched.

Although we have not performed a sufficient number of runs (all design optimization runs with CFD calculations even if meta-models or surrogate models are employed are computationally expensive) to give a clear preference to the DMFFD, it seems that DMFFD will give us more flexibility in the optimization and it will also allow to make design changes more clearly visible to the engineer during the optimization.

References

Arima, T., Sonoda, T., Shirotori, M., Tamura, A., and Kikuchi, K. (1999). A numerical investigation of transonic axial compressor rotor flow using a low-Reynolds-number k-ε turbulence model. *ASME Journal of Turbomachinery*, 121(1), pp. 44–58.

Bihrer, T. (2006). Direct Manipulation of Free-Form Deformation in Evolutionary Optimisation. Diploma Thesis. Computer Science Department/ Simulation, Systems Optimization and Robotics Group (SIM), TU Darmstadt, Germany

Coquillart, S. (1990). Extended Free-Form Deformation: A Sculpturing Tool for 3D Geometric Modeling. *Computer Graphics*, 24(4):187–196.

Fogel, D.B. (1995). Evolutionary Computation: toward a new philosophy of machine learning New York, NY: IEEE Press.

Foli, K., Okabe, T., Olhofer, M., and Sendhoff, B. (2006). Optimization of Micro Heat Exchanger: CFD, Analytical Approach and Multi-Objective Evolutionary Algorithms. *International Journal of Heat and Mass Transfer*, 49(5-6), pp. 1090–1099.

Geist, A. et al. (1995). PVM: Parallel Virtual Machine – A Users' Guide and Tutorial for Networked Parallel Computing. Cambridge, Ma: The MIT Press.

Hansen, N., and Ostermeier, A. (2001). Completely Derandomized Self-adaptation in Evolution Strategies. In: *Evolutionary Computation*, vol. 9, no. 2, pp. 159–196.

Hasenjäger, M., Sendhoff, B., Sonoda, T., and Arima, T. (2005). Three dimensional evolutionary aerodynamic design optimization with CMA-ES. In: *GECCO 2005: Proceedings of the 2005 Conference on Genetic and Evolutionary Computation*, edited by H.-G.~Beyer et al. ACM Press, New York, NY, pp. 2173–2180.

Hasenjäger, M., Sendhoff, B., Sonoda, T., and Arima, T. (2005). Single and Multi-Objective Approaches to 3D Evolutionary Aerodynamic Design Optimisation. In: *Proceedings of the 6th World Congress on Structural and Multidisciplinary Optimization*, Rio de Janeiro, Brazil.

Hsu, W.M., Hughes, J.F., and Kaufman, H. (1992). Direct Manipulations of Free-Form Deformations. *Computer Graphics*, 26:177–184.

Kanazaki, M., Obayashi, S., Morikawa, M., and Nakahash, K. (2002). Multiobjective Design Optimization of Merging Configuration for an Exhaust Manifold of a Car Engine. In: *Parallel Problem Solving from Nature - PPSN VII*, Guervos, Admidis, Beyer, Fernandez-Villacanas and Schwefel (Eds.), Lecture Notes in Computer Science, 2439, pp. 281–287.

Lépine, J., Guibault, F., Trépanier, J.Y., and Pépin, F. (2001). Optimized Nonuniform Rational B-spline Geometrical Representation for Aerodynamic Design of Wings. *AIAA Journal*, Vol. 39, No. 11.

Menzel, S., Olhofer, M., and Sendhoff, B. (2005). Evolutionary Design Optimisation of Complex Systems integrating Fluent for parallel Flow Evaluation. In: *Proceedings of European Automotive CFD Conference*, pp. 279–289.

Olhofer, M., Jin, Y., and Sendhoff, B. (2001). Adaptive encoding for aerodynamic shape optimization using Evolution Strategies. In: *Congress on Evolutionary Computation (CEC)*. IEEE Press, Seoul, Korea, pp. 576–583.

Ong, Y.S., Nair, P.B., and Lum, K.Y. (2006). Max-Min Surrogate Assisted Evolutionary Algorithm for Robust Aerodynamic Design. *IEEE Transactions on Evolutionary Computation*, 10(4), pp. 392–404.

Oyama, A., Obayashi, S., and Nakamura, T. (2000). Real-coded Adaptive Range Genetic Algorithm Applied to Transonic Wing Optimization. *Proceedings of the 6th International Conference on Parallel Problem Solving from Nature*. Springer, pp. 712–721.

Perry, E.C., Benzley, S.E., Landon, M., and Johnson, R. (2000). Shape Optimization of Fluid Flow Systems. In: *Proceedings of ASME FEDSM'00. ASME Fluids Engineering Summer Conference*, Boston.

Piegl, L., and Tiller, W. (1997). The NURBS Book. Springer-Verlag Berlin Heidelberg.

PVM: Parallel Virtual Machine. http://www.csm.ornl.gov/pvm/pvm_home. html.

Rechenberg, I. (1994). Evolutionsstrategie '94. Friedrich Frommann Verlag· Günther Holzboog, Stuttgart-Bad Cannstatt.

Requicha, A.A.G. (1980). Representations for Rigid Solids: Theory, Methods, and Systems. In: *Computing Surveys* 12, pp. 437–464.

Schwefel, H.P. (1995). Evolution and Optimum Seeking. John Wiley & sons. New York.

Sederberg, T.W., and Parry, S.R. (1986). Free-Form Deformation of Solid Geometric Models. *Computer Graphics*, 20(4):151–160.

Sendhoff, B., Kreuz, M., and von Seelen, W. (1997). A condition for the genotype-phenotype mapping: Causality. In: *Thomas Bäck, editor, Proceedings of the Seventh International Conference on Genetic Algorithms (ICGA'97)*, Morgan Kauffmann, pp. 73–80.

Sonoda, T., Yamaguchi, Y., Arima, T., Olhofer, M., Sendhoff, B., and Schreiber, H-A. (2004). Advanced High Turning Compressor Airfoils for Low Reynolds Number Condition, Part 1: Design and Optimization. *Journal of Turbomachinery*, 126: 350–359.

Terzopoulos, D., and Qin, H. (1994). Dynamic NURBS with geometric constraints for interactive sculpting. In: *ACM Transactions on Graphics* 13, Nr. 2, pp. 103–136.

Ugail, H., and Wilson, M.J. (2003). Efficient shape parameterisation for automatic design optimisation using a partial differential equation formulation. In: *Computers & Structures* 81, pp. 2601–2609.

Chapter 5

EVOLVING MICROSTRUCTURED OPTICAL FIBRES

Steven Manos[1] and Peter J. Bentley[2]

[1] *Optical Fibre Technology Centre, University of Sydney, Australia, and Centre for Computational Science, Department of Chemistry, University College London;* [2] *Department of Computer Science, University College London, Gower St, London WC1E 6BT.*

Abstract   Optical fibres are not only one of the major components of modern optical communications systems, but are also used in other areas such as sensing, medicine and optical filtering. Silica microstructured optical fibres are a type of optical fibre where microscopic holes within the fibre result in highly tailorable optical properties, which are not possible in traditional fibres. Microstructured fibres manufactured from polymer, instead of silica, are a relatively recent development in optical fibre technology, and support a wide variety of microstructure fibre geometries, when compared to the more commonly used silica. In order to meet the automated design requirements for such complex fibres, a representation was developed which can describe radially symmetric microstructured fibres of different complexities; from simple hexagonal designs with very few holes, to large arrays of hundreds of holes. This chapter presents a genetic algorithm which uses an embryogeny representation, or a growth phase, to convert a design from its genetic encoding (genotype) to the microstructured fibre (phenotype). The work demonstrates the application of variable-complexity, evolutionary design approaches to photonic design. The inclusion of real-world constraints within the embryogeny aids in the manufacture of designs, resulting in the physical construction and experimental characterisation of both single-mode and high-bandwidth multi-mode microstructured fibres, where some GA-designed fibres are currently being patented.

Keywords: microstructured optical fibres, evolutionary design, genetic algorithm, embryogeny

S. Manos and Peter J. Bentley: *Evolving Microstructured Optical Fibres*, Studies in Computational Intelligence (SCI) **88**, 87–124 (2008)
www.springerlink.com

1.   INTRODUCTION

Whether used for telecommunications, cable television or in the operating theatre, optical fibres are one of the major components used in modern optical communications systems. Conventional optical fibres are typically made of silica, and have a refractive index profile that varies across the fibre but is constant along the fibre. In solid-core silica fibres, the refractive index profile is altered by the inclusion of chemicals, such as fluoride, germanium and boron, into the silica preform, which is several centimeters in diameter. This preform is then heated and stretched down to an optical fibre, typically $100\text{--}200\,\mu m$ in diameter.

The development of silica microstructured optical fibres has expanded the application areas of optical fibres. The inclusion of air holes results in highly tailorable optical properties not possible in traditional silica fibres. Silica microstructured fibres primarily have hexagonal arrays of holes, a result of the capillary stacking approach used to manufacture preforms. Microstructured polymer optical fibres (MPOF) are a more recent advance which allows almost arbitrary symmetry and positioning of holes, opening up the design space considerably. To reflect this, new search and optimisation algorithms need to be developed which can generate the required diversity of designs and, in parallel, deal with manufacturing constraints.

This chapter presents a genetic algorithm which uses an embryogeny, or a growth phase to convert a design from its genetic encoding (genotype) to the microstructured fibre (phenotype). The result is a compact binary representation which can generate microstructured fibre designs, where the number of holes and the overall complexity of the designs is not predetermined, but can evolve over time. A brief overview of MPOFs and their manufacture is presented in Section 1.1, followed by a discussion of previous representations for microstructured fibre optimisation in 1.2, and motivating the need for more sophisticated design algorithms. Representations and embryogenies in genetic algorithms are then introduced in Section 1.3. Details of the new representation and embryogeny for microstructured fibres are given in Sections 2.1 and 2.2, with a discussion of symmetry in Section 2.3. The genetic operators mutation and recombination are discussed in Section 2.4, followed by an overview of the population selection operators, described in Section 2.5. The simple fitness function used to generate the results presented in this chapter is given in 2.6.

Section 3 outlines a test case for the algorithm, where microstructured fibres were designed using the simple fitness function that approximates bandwidth. Other examples of using the GA for practical microstructured fibre design are discussed in Section 4. Conclusions are given in Section 5.

1.1 Microstructured Optical Fibres

Microstructured optical fibres (MOF) are fibres which use air channels that run the length of the fibre to guide light. The first MOF was reported by Knight *et al* (1996), and consisted of a hexagonal array of air holes. A defect in the core was formed by excluding the central air hole in the lattice, forming a region where light could be guided. Since then, various aspects of MOFs, from theory to manufacture, have been very active areas of research. The use of air holes means that much larger index contrasts are possible than with chemical doping. Propagating modes experience this high index contrast, and in combination with more freedom in the types of designs possible due to the 2-dimensional profile, MOFs can provide highly tailorable optical characteristics. The first MOFs were 'endlessly single-mode fibres' (Birks *et al* 1997), achieving single-mode guidance over the entire spectrum. MOFs have been used to produce super continuum generation (Ranka *et al* 2000), where a dispersion zero at the pump laser wavelength is employed to generate a continuous range of wavelengths (white light) from a single wavelength laser source. They have also been developed for more traditional applications, such as dispersion flattened fibres (Renversez *et al* 2003). As more optical effects are required, MOF designs tend to have a variety of hole sizes, making the designs more complex.

Silica MOFs are usually manufactured by stacking silica tubes into a preform, which naturally pack into a hexagonal array. During the fibre drawing process, the geometry of the fibre can be manipulated by altering drawing conditions such as pressurisation, temperature and draw speed, resulting in, for example, fibres with non-circular holes, or large air fractions. As the air inclusions in MOFs have a lower refractive index than the core material, light typically leaks through the gaps between the holes, and can also tunnel through holes. Although this leakage is an intrinsic feature of MOFs, the positions and sizes of holes can be tailored to guide light with negligible signal loss. Light guidance in MOFs can be primarily explained using the average index approximation: the region of holes (referred to as the cladding) around the central solid core is on average seen as a lower refractive index, confining light to the core region (Figure 5-2b,c).

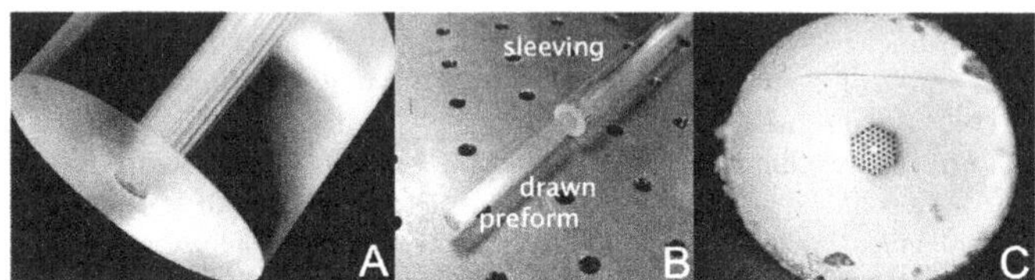

Figure 5-1. Three steps used in MPOF manufacture. Holes are drilled into a PMMA preform 75 mm in diameter and 70 mm long (A). This is drawn down to an intermediate cane (B) which can be sleeved with a PMMA tube to allow further reduction in the size of the microstructured region. This cane is then drawn to an optical fibre (C), 200 um to 1000 um in diameter.

Table 5-1. Air hole encoding and corresponding drill bit and fibre hole sizes, for a reduction factor of 340.

r_i value	Preform drill hole diameter (mm)	Fibre hole diameter (μm)
0	1.2	3.5
1	1.5	4.4
2	2.0	5.9
3	3.0	8.8
4	4.5	13.2
5	6.0	17.6
6	7.0	20.6
7	8.5	25.0

Polymer optical fibres (POF) have long been used where properties such as large cores and high data rates are desired. The use of polymer is advantageous for several reasons. Due to its lower elastic modulus compared to silica, the manufacture of large core fibres up to 1 mm or more in diameter is possible, while maintaining mechanical flexibility and mechanical stability. As a result, they are easier to handle, and polymer fibres are also generally cheaper to produce than silica fibers. One of the drawbacks of polymer is the intrinsic loss associated with absorption and microscopic scattering. However, this issue is less critical for short distance applications in the order of 10 to 100 meters, and as a result, this is the main application base for POF.

A more recent development which has brought together the advantages of MOF and POF technologies has been the manufacture of Microstructured Polymer Optical Fibres (MPOF) (van Eijkelenborg *et al* 2001a). Since polymer preforms can not only be stacked like silica, but also drilled, cast and extruded, this results in a wide range of possible fibre designs. One of the major advantages is that the arrays of air holes are no longer limited to hexagonal structures, and many other geometries are now possible. Improvements in fabrication techniques have also led to lower losses (van Eijkelenborg *et al* 2004, Large *et al* 2006).

Today the manufacturing method most often used to prepare preforms at the Optical Fibre Technology Centre is drilling. Holes are drilled into a poly-methyl methacrylate (PMMA) preform approximately 75 mm in diameter and 70 mm long using a computer controlled mill (Figure 5-1A). A limited set of drills are available, resulting in a discrete set of hole sizes (Table 5-1). These results in the first important manufacturing constraint - that only particular hole sizes are available. The second manufacturing constraint is the minimum distance between holes when the preform is drilled, which is required to help prevent wall collapse or fracturing. Typically 0.2 to 0.5 mm is sufficient. The preform is then heated, and drawn to a cane, usually 5 to 10 mm in diameter (Figure 5-1B). This can then be sleeved with more PMMA, and then undergoes a further draw to produce an optical fibre of diameter 200 μm to 1000 μm in diameter (Figure 5-1C). This results in a size reduction ranging from 300 to 1000 times, compared to the original preform. The second column in Table 5-1 shows the corresponding fibre hole sizes for a reduction of 340. Holes can change both their relative size and shape during drawing (Xue *et al* 2006). These deformations depend on the draw conditions (temperature for example), but also depend on the proximity and size of surrounding holes. For simplicity however, in this work we presume that holes are undeformed during the draw process.

1.2 Previous Representations for Microstructured Fibre Design

The design of microstructured fibres has to date mainly been a human-driven process of analysis, where optical properties such as single-modedness, high-nonlinearity and dispersion are considered. Examples of optimisation in the literature have mainly focused on simple structures, see Figure 5-2. Extensions of this design space to multiple rings with different sized holes have also been published (Poletti *et al* 2005), with fitness functions involving more complex optical properties, such as a constant dispersion over a range of wavelengths (dispersion flattening). Since the designs can be described in terms of a simple parameterisation, it is straightforward to envisage all possible designs. Allowing both the symmetry and hole properties to change, however, increases the range of possible designs.

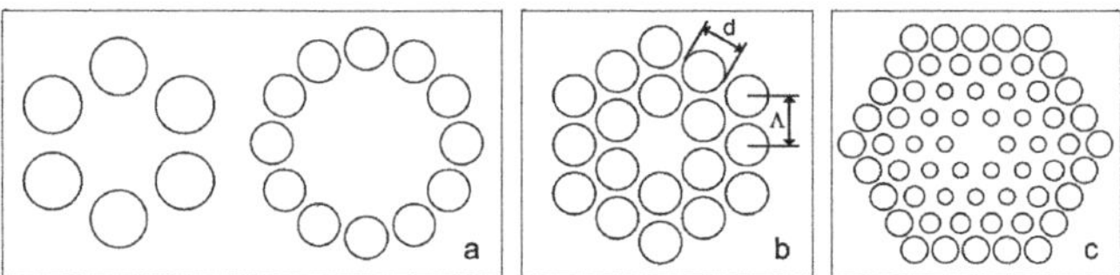

Figure 5-2.   Various types of microstructured optical fibre profiles that have been used for EA optimisation in the literature. Single rings of holes (a) were used in van Eijkelenborg et al (2001). Standard hexagonal arrays using a fixed number of rings (b) were used in Manos et al (2002). More complicated hexagonal arrays (c) resulted in better dispersion performance in Poletti et al (2005), where up to 6 design parameters were used to represent the microstructured fibre.

With the development of new manufacturing techniques to produce Microstructured Polymer Optical Fibres (MPOF), designs of almost arbitrary hole arrangements are possible.

For example, Figure 5-3E is a simple square hole array MPOF, and Figure 5-3F is a far more complicated MPOF with 171 holes using 4 different hole sizes. Given this flexibility, more interesting design questions can be posed.

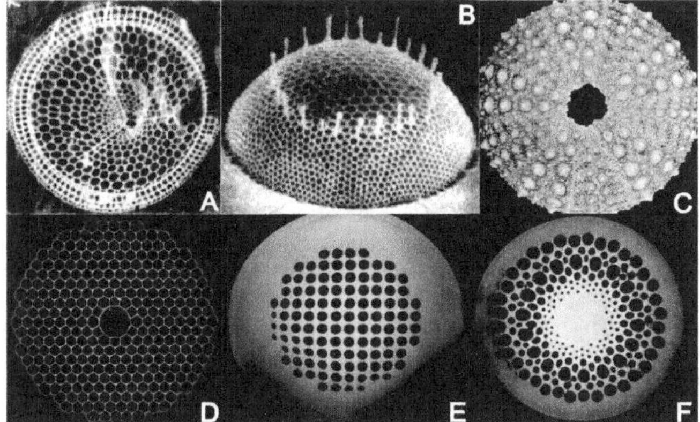

Figure 5-3. Examples of symmetric structures found in nature and microstructured optical fibre structures. A and B are images of radiolarians, microscopic ocean-dwelling microbes whose skeletal structure is made from a silica-like material. C is the common sea urchin, which exhibits 5-fold radial symmetry. D is an example of a silica microstructure fibre (or a photonic crystal fibre), where the hexagonal pattern can clearly be seen. E and F are polymer microstructured fibres. E is a square array of holes which can be used for medical imaging, and F is a more complex fibre design with large variations in hole positions and hole sizes. These sorts of structures are more easily made using polymer-based fibres.

Rather than focusing on optimising the sizes of holes in a hexagonal array with a fixed number of rings, one can ask "what is the best symmetry for this sort of design?", or "can we achieve these properties with fewer holes?". Given the huge design space of symmetries and hole arrangements that can be explored, an automated design algorithm would need to be able to evolve designs which can change symmetry and their complexity (for example, overall number of holes) over time. A genetic algorithm is implemented to evolve these designs.

1.3   Representations

The genetic representation is fundamental to an evolutionary algorithm, and used to store details of the design. Simple representations are generally fixed in length, resulting in designs which are variations of a pre-defined design type. However, appropriate representations can be developed which encode the topology, complexity and other large-scale features, allowing the overall design to change over time.

1.3.1   Embryogenies.   Typically in evolutionary algorithms the conversion from genotype to phenotype is quite direct, for example converting a binary genotype to numerical values. An embryogeny refers to a conversion where the genotype is used as an encoded set of 'instructions,' which are used to grow or develop the phenotype. This is a characteristic of developmental systems found in nature, resulting in the complexity of the phenotype being greater than the underlying genotype. Some examples of symmetric phenotypes found in nature are shown in Figures 5-3A-C. Two of the main features of these sorts of systems are:

- Indirect correspondence between features of the genotype and phenotypic features. The genotype acts as a set of instructions to grow or develop designs;
- Polygeny. Phenotypic characteristics of designs are not dependent on just one part of the genotype, but can arise from the effects of multiple genes acting in combination.

Embryogenies offer many advantages over simple representations for the design of complex objects such as microstructured fibres. Using a direct representation, the location and size of every single hole in a microstructured fibre would have to be described. As more holes are defined in the genotype, the search becomes less efficient, especially since many of the genes will interact when considering manufacturing constraints, such as the minimum allowed distance between holes.

Embryogenies can also exploit hierarchy to re-use parts of the genotype, for example, sets of holes which reappear at different locations in the structure. This results in a more efficient and lower-dimensional search space. Symmetry is particularly relevant in microstructured fibre design since rotational symmetry, or repeating patterns, are common in fibre designs. If one imagines all the symmetries and hole arrangements possible for microstructured fibres, the phenotypic space is massive.

In converting from the compact genotype to extended phenotype, embryogenies employ a process, or a set of instructions. These instructions can be intrinsic to the representation, where they are defined within the genotype. They can also be extrinsic or even external (Kumar and Bentley 1999), where an external algorithm is used as a set of instructions to develop the phenotype according to values defined by the genotype.

Although embryogenies are powerful representations for complex designs, there are drawbacks. Embryogenies, along with the corresponding mutation and recombination operators, are often hand designed, since there are no set rules or methodologies on how to design an embryogeny. The representations themselves can also evolve (Stanley and Miikkulainen 2003).

Since the conversion from genotype to phenotype is now a process, rules can be built into this algorithm that can ensure that the genotype will produce valid designs. The embryogeny is important here, since in the case of a more direct representation, the interactions between the multiple genes would result in many complex constraints that would be difficult or impossible to satisfy. For example, a direct representation encoding n holes would require the simultaneous satisfaction of $n(n-1)/2$ combinations of constraints, adding to the complexity of the search space.

2.   A REPRESENTATION FOR MICROSTRUCTURED OPTICAL FIBRES

Representations express how a phenotype is encoded as a genotype. The representation consists of two major parts: the genotype, which encodes the details associated with the design, and the embryogeny, which is the algorithm that converts from genotype to phenotype. The genotype used here defines two major features of the microstructured fibre, the positions and sizes of a variable number of air holes, and the rotational symmetry of the structure. The embryogeny is then used to convert the genotype into a valid microstructured fibre design, ensuring that manufacturing constraints such as non-overlapping holes, are maintained. In developing the representation, two main factors were considered:

1. Exploitation of radial symmetry to minimise the length of the genotype, but simultaneously ensure the design of symmetric microstructured fibres,

2. The direct incorporation of constraints into the embryogeny, for example, discrete, rather than continuous sizes of holes, and other manufacturing constraints such as the minimum allowed distance between holes.

Each part of the representation will now be explored in detail.

2.1   Genotype

The genotype is used to store

- An integer value which represents the global rotational symmetry of the microstructured fibre n_{symm},

- The position and size of each hole x_i, y_i, r_i where $i \in [1, \ldots, n_h]$ and n_h is the number of holes in a sector.

Only the details of holes in one repeating unit of the designs are stored, so that the total number of holes encoded by the genotype is $n_h \times n_{symm}$. The fibres considered here are made out of polymer, and the holes all contain air, so no information about the materials needs to be explicitly defined in the genotype (although this can be added for more complex designs). The position values x_i, y_i are decimal values encoded in binary form. r_i is an integer which represents a discrete hole size due to the limited availability of hole sizes (Table 5-1) used in the drilling preparation of the MPOF preform. If need be, direct decimal values can be used instead. Symmetries based on reflection are not included, but could be added later.

The above values are individually represented in binary form, and concatenated into a single binary string. Thus, existing mutation and recombination operators for the cutting and splicing of binary strings can be used (Goldberg 1989). A non-binary genotype which uses a mix of integers and real-values could be implemented, but has not been done as it would require unnecessary complexity in the recombination and mutation operations.

The bit length of n_{symm} is b_s, the x_i, y_i values have equal length $b_{x,y}$ and the hole size value has length b_r. The total genotype length b_t, is

$$b_t = b_s + n_h(2b_{x,y} + b_r) \tag{1}$$

This relationship encapsulates the variable length of the genotype. Holes which are represented by a binary string of length $2b_{x,y} + b_r$ can be added to or deleted from the genotype, whilst still maintaining a length which satisfies the above equation.

2.2 Embryogeny

The conversion of the binary genotype to a valid phenotype involves four steps, the decoding of the binary genotype to its numerical values, and then the development of the radially symmetrical design.

Step 1: Decoding the Binary Genotype

The first phase of the genotype to phenotype mapping involves decoding the genotype into the numerical values it represents. The first b_s bits are decoded into an integer representing the symmetry n_{symm}.

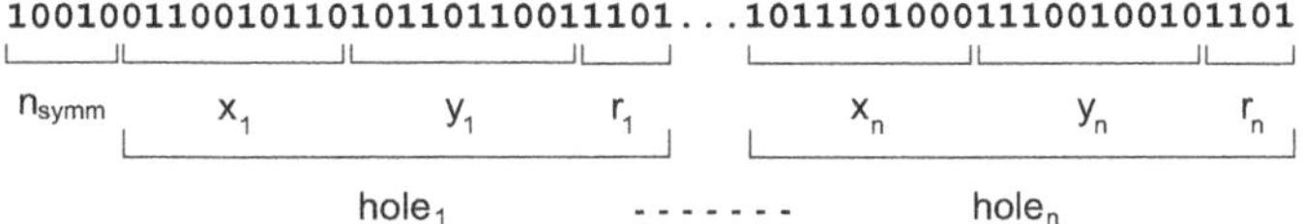

A different number of bits can be used to represent different upper limits on the symmetry. A typical value used is $b_s = 5$ resulting in a maximum symmetry of $n_{symm} = 2b_s - 1 = 31$. Symmetry is discussed further in Section 2.3.

The n_h holes are then decoded into their positions and hole sizes. The Cartesian position values x_i, y_i are decoded into unsigned decimal values, thus lying in the first quadrant and easily generating an $n_{symm} = 4$ symmetry. As an example, if $b_{x,y} = 9$, where the first 6 bits represent the whole component and the remaining 3 bits represent the fractional component, this would result in a minimum increment of $0.125\,\mu$m and a maximum value of $63.875\,\mu$m. Larger x_i, y_i values can be achieved by increasing $b_{x,y}$ or by scaling the values of x_i, y_i.

The mapping of the r_i values to hole sizes is treated a little differently. Since only particular hole sizes (drill bits) are available for use, the r_i binary string is converted into an integer value. The different integer values map to the hole sizes which are available. A choice of $b_r = 3$ specifies 8 different hole sizes. Table 5-1 shows the available spread of drill bits and resulting fibre hole diameters. These decoded positions form the basic $n_{symm} = 4$ structure, an example of which is shown in Figure 5-4, Step 1.

Steps 2 and 3: Converting to a Symmetrical Fibre

The basic decoded structure initially corresponds to the correct $n_{symm} = 4$ (Figure 5-4, Step 1). The next step is to transform this structure symmetry. The polar r_i, θ_i coordinates are evaluated for each hole, and then the angles are transformed as

$$\theta_i = \frac{4}{n_{\text{symm}}}\theta_i, \ i \in [1, \ldots, n_h] \tag{2}$$

An example is shown in Step 2 in Figure 5-4 for $n_{symm} = 7$. This group of holes, referred to as the unit holes, are then copied radially $n_{symm} - 1$ times in steps of $\theta = 2\pi/n_{symm}$, resulting in the complete symmetrical microstructured fibre as shown in Step 3.

Step 4: Growing the Microstructured Fibre

From Figure 5-4, Step 3, the holes are initially overlapping or touching after being converted into a complete symmetrical structure. This final stage of the embryogeny involves 'growing' the size (radii) of the holes in order to obey manufacturing constraints. More specifically, each hole must be surrounded by a minimum wall thickness w_h for structural stability.

The two main ideas behind the described algorithm are:

- The algorithm steps through the available hole sizes, such as those in Table 5-1, until holes either equal their own maximum defined size r_i, or begin to overlap with neighbouring holes,
- When holes stop growing, the growth of corresponding holes in the other symmetrical units is also stopped in parallel to ensure a symmetrical fibre.

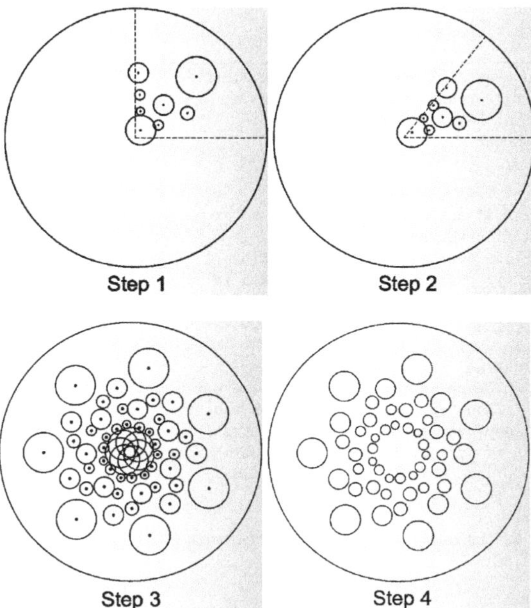

Figure 5-4. Genotype to phenotype conversion of a microstructured fibre as outlined in the text. The fibre immediately after decoding from the binary genotype is shown in Step 1. Step 2 is the initial transformation into the fibre's new symmetry nsymm $= 7$. Step 3 is the result of copying the symmetry unit from Step 2 rotationally around the fibre. Finally, Step 4 shows the result of growing the hole radii in order to satisfy hole overlapping constraints.

This results in Figure 5-4, Step 4, with all holes at least w_h apart. Some of the holes have achieved their genotype size, whereas others are smaller due to the presence of surrounding holes. The holes in the centre have not been expressed phenotypically at all because of their close proximity to one another. The growth process which results in the Step 4 design can be seen as a 'correction' to the phenotype from Step 3 (Yu and Bentley 1998), forming a legal mapping. Since all individuals are phenotypically correct with respect to constraints, the overall performance of the GA is high since no effort is wasted on infeasible solutions that must be removed through penalty approaches or other methods. (The algorithm is outlined in detail in Figure 5-5.)

```
loop over list of available hole sizes {
  get 'currentHoleSize' from list of available hole sizes
  if no holes are growing {
    break
  } else {

    // If any holes are allowed to grow, set their size to the
    // currentHoleSize
    loop over all holes in the fibre {
      if current fibre hole growth is 'true' {
      set current fibre hole radius to 'currentHoleSize'
      }
    }
    define list holesToStoprLimit;
    // Check to see if any holes have reached their own
    // genotype defined limits on radius
    loop over all holes in the fibre {
      if current fibre hole radius==genotype-defined radius {
        add hole to holesToStoprLimit;
      }
    }
    define lost holesToStopOverlapLimit;
```

```
// Do a pairwise check of all holes to check if any
// overlap
loop over all unique pairwise holes [H1,H2] in the fibre {
  if holes H1,H2 overlap {
     add hole H1 to holesToStopOverlapLimit
     add hole H2 to holesToStopOverlapLimit
  }
}
// Note that now the rules are applied in parallel
// to maintain symmetry
loop over all holes in holesToStoprLimit {
  if hole growth is set to true {
    set hole growth to 'false'
    if using the first (smallest) currentHoleSize {
      set hole radius to 0
    } else {
      set hole radius to previous holeRadius
      }
  }
}
loop over all holes in holesToStopOverlapLimit {
  if hole growth is set to true {
  set hole growth to false
    }
  }
}
}
```

Figure 5-5. Algorithm used to 'grow' the microstructured fibre designs, which corresponds to the transition from Step 3 to Step 4 in Figure 5-4.

The phenotypic hole sizes are not propagated back into the genotype. The genotype serves the purpose of specifying a 'potential' hole size, then the surrounding phenotypic features and the genotype itself govern the extent to which holes can grow. Thus, any genotype, even a random string of binary values, is a feasible design. A number of randomly generated designs are shown in Figure 5-6.

One phenotypic feature that cannot occur as a result of the algorithm is a central air hole. This can be included by adding a binary string of length br, representing the size of a hole at the origin. The size of the hole would be influenced by the surrounding holes, or a decoded size value of 0 could specify a zero hole radius. The symmetry copying operation would not be applied to this hole, thus the global n_{symm} symmetry would still be maintained.

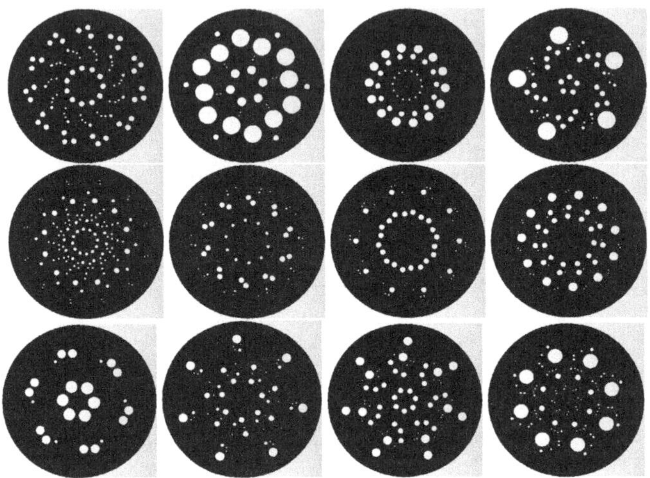

Figure 5-6.   Phenotypes corresponding to randomly generated binary genotypes, using the fixed hole sizes outlined in Table 5-1. Each microstructured optical fibre design varies in symmetry, overall size and the number of holes.

2.3   Symmetry

Figure 5-7 shows an example of a design where the binary part of the genotype describing the holes x_i, y_i, r_i remain constant, but the symmetry is varied $n_{symm} = 2, 3, \ldots, 10$. Designs with $n_{symm} = 0$ could be used to represent circularly symmetric fibres, and $n_{symm} = 1$ for random arrangements of holes. However, neither of these two cases are of interest in the real problems discussed here, thus, values of 0 or 1 are mapped to $n_{symm} = 2$. Limiting the representation to designs which exhibit rotational symmetry of at least 2 is important for various reasons. The first is related to the computational evaluation of the properties of fibres. The evaluation of modes in random structures is the most computationally demanding case. Any symmetry in the design can be exploited to reduce the computational effort required, such that only a slice of angular width $2\pi/n_{symm}$ needs to be evaluated (Poladian *et al* 2002, Issa and Poladian 2003).

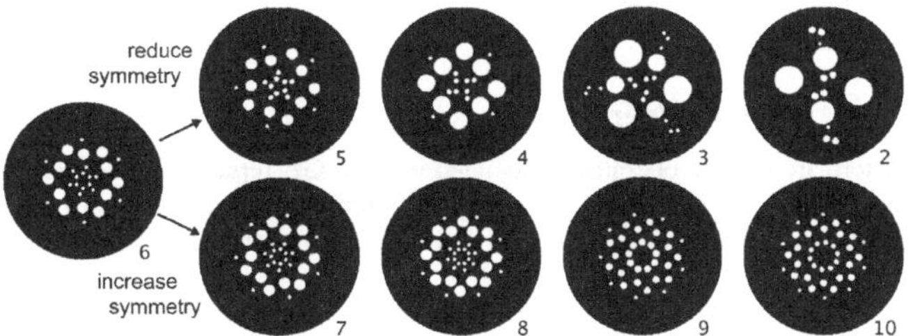

Figure 5-7. An optical fibre which has been converted to various symmetries. On the top row the symmetry is reduced from 6 to 2, and on the bottom it is increased from 6 to 10 (note that higher symmetries are possible). The influence of the growth phase is evident here with the reduction of hole sizes as the distance between neighbouring holes decreases as the symmetry is increased. Some holes disappear completely, whereas others maintain the same size through most symmetries. Overall the same design 'theme' is maintained, and the growth phase alters the hole sizes to maintain appropriate constraints from symmetry to symmetry.

Secondly, symmetry is important in the drawing of the fibre. Symmetry is irrelevant in the manufacture of the actual preform, since essentially any arrangement of holes can be accommodated when drilling is used. It becomes particularly important, however, during the drawing process. Hole deformations occur, and the types of and extent of deformations relate strongly to the surrounding holes (Xue *et al* 2006). Any digression from asymmetric pattern of holes can cause larger deformations to occur.

Thirdly, symmetry can help control and tune optical effects. Designs with random hole arrangements have been published (Monro *et al* 2000, Kominsky *et al* 2003), where the optical effects achieved in these fibres are typically a result of homogenization effects which are relatively robust with respect to perturbations introduced during manufacture. Not all optical properties, however, are robust to such perturbations. The dispersion over large wavelength ranges, and single-moded guidance, for example, have been more effectively controlled and tuned using symmetrical structures in the literature. Further, features of symmetrical structures can be characterised easily, such as the spacing between holes in a hexagonal array. This results in the ability to more easily express the relationships between phenotypic features optical properties.

Finally, using symmetry to repeat the defined patterns of holes reduces the dimensionality of the search space.

2.4   Genetic Operators

There are genetic operators associated with all GA representations. For example, recombination of genotypes from different individuals is used to create new designs, and mutation is used to modify existing genotypes.

2.4.1   Initialisation.   Initialisation involves the generation of binary strings of length $b_s + n_h \left(2b_{x,y} + b_r\right)$ to seed the population. For each individual, a random value of n_h is used, typically $n_h \in [1, 15]$, to generate individuals with different numbers of holes. The genotype of length b_t is then filled with random binary values. In some cases pre-defined fibres can be used to seed the population, but generally the GA is seeded using a completely random population (for example, the designs in Figure 5-5).

2.4.2   Recombination.   In the work discussed here, the population consists of a group of designs which vary in the genotype length, and the recombination operator must deal with recombining individuals of different lengths and also producing children with valid genotype lengths (Equation 1). The result is that limitations are placed on where splicing takes place during variable length crossover to maintain valid genotypes.

Single point crossover between two parents is used to produce one or two children. The breeding pairs are typically chosen using $k = 2$ tournament selection, denoted parent 1 and parent 2.

Given parent 1 with binary length $b_{t,1}$, a random splice point is chosen in the range $[1, b_{t,1}]$. This splice point can sit within n_{symm} on the edge of binary strings representing holes or within those binary strings themselves. An example with 3 holes, where the splice point occurs within x of the 3rd hole is

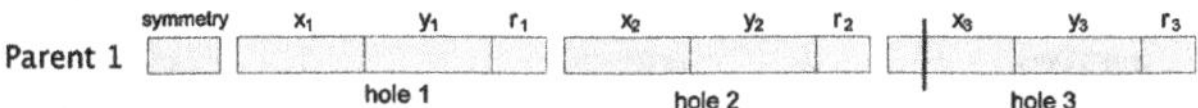

In order for the child's genotype to have a valid length, the recombination point on the 2nd parent must be chosen at the same point within the n_{symm} or x_i, r_i, r_i binary substring. Given parent 2 with $n_{h,2}$ holes, a random integer in the range $[1, n_{h,2}]$ is generated which picks which hole the splice will occur.

The exact cut position within the parent 2 binary string depends on the cut point selected for parent 1 in order to maintain the correct length. Once these splice points have been defined, one, or two children, can be generated by joining the light grey binary strings (parent 1) and the dark grey strings (parent 2).

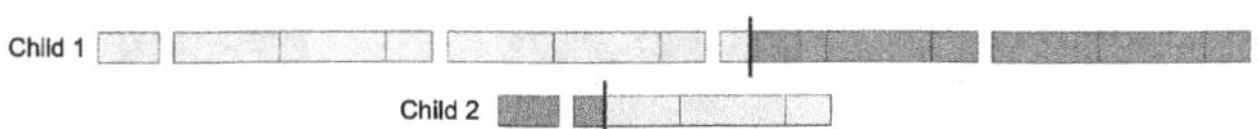

From two parents with 3 and 2 holes, children with 4 holes and a single hole were generated. Thus, the complexity of designs is able to change. Typically the GA only retains one of the two children. In the case where pure elitist survival selection is used, more parents have the opportunity to breed and contribute genetic material to the child population, resulting in more diverse child population. Examples of the phenotypes of recombined individuals are shown in

Figure 5-8. Although the genotype and corresponding recombination operators are very simple, the resulting phenotypic behaviour of individuals that are
recombined is quite complex.

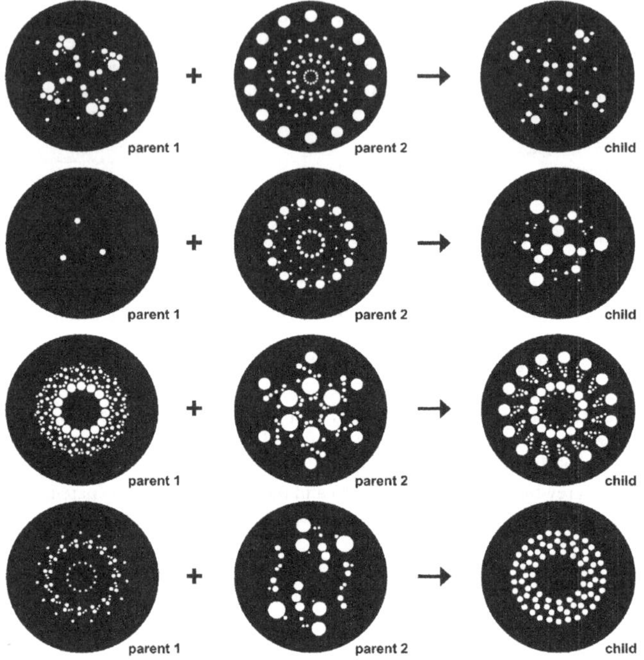

Figure 5-8.   Example of single child recombination from two randomly generated parents with
different lengths. No mutation has been used. The inheritance of features such as symmetry,
groups of holes and single holes can be seen. For example, in the first recombination example
a child with nsymm $= 4$ is formed which contains holes from both parents. In the second
example Parent 1 is acting as a symmetry operator, turning the child into a copy of Parent 2 but
with nsymm $= 3$ and the single holes from Parent 1. Characteristics from both parents are also
evident in the final two examples.

2.4.3   Mutation.   Two types of mutation operators are used. One operates on individual bits, and other operates on whole substrings at a time (Figure 5-9).

Different mutation rates are used on different parts of the genotype. A mutation rate p_s is used for the bits representing the symmetry value. The rate $p_{x,y}$ is used for the bits representing the position values, and the rate p_r for the hole size values. Higher level hierarchical operators are used which can add and delete complete holes from the genotype, increasing or decreasing the total number of holes. The mutation rate p_a is associated with the rate of hole addition, where a random string of length $2b_{x,y} + b_r$ is spliced onto the genotype. Binary segments can also be deleted at a rate of p_d, where a random hole is chosen for deletion. Since single point crossover tends to recombine holes which lie far apart within the genotype, another operator randomly selects two holes in a genotype and swaps their positions. This occurs at a rate p_o. Note that changing the ordering of holes in the genotype has no effect on the phenotype. Examples of mutation on various designs are shown in Figure 5-10.

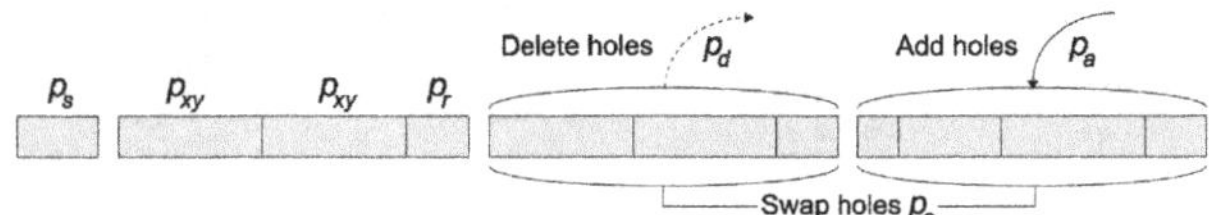

Figure 5-9.   Schematic description of the two classes of mutations used on the variable length binary genotype.

2.5   Population Operators

Population operators include selection for survival and selection for breeding. In selecting which individuals survive from one generation to the next, a pure elitist scheme is used, such that both parents and children have the opportunity to survive. The population of size $N_p + N_c$ is ordered from best to worse in terms of the individual fitness values, and the top N_p designs survive. In the case of multiple objectives, the NSGA-II Pareto ranking and distance measures (Deb *et al* 2000) are used to completely order the population.

In choosing individuals for breeding, $k = 2$ tournament selection is used. Since the breeding of two parents results in a single child, $2N_c$ tournaments are played to breed N_c children.

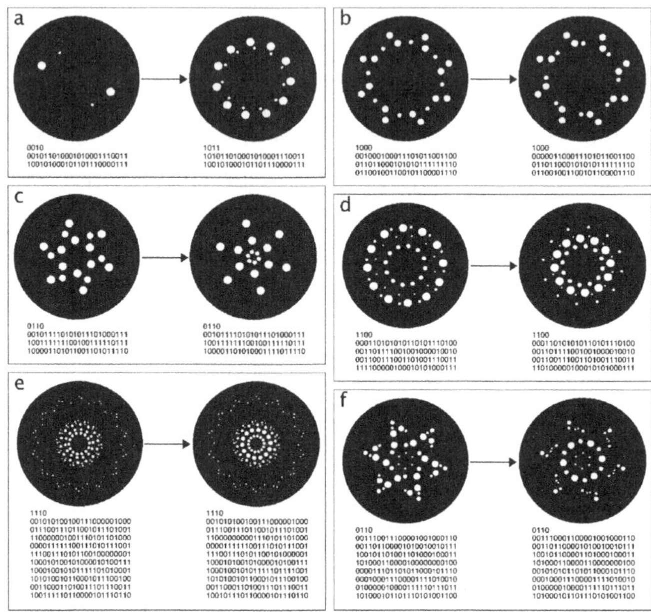

Figure 5-10. Six examples of randomly generated fibre design (left) and a random mutation (right), along with the binary representation of each design. The first line below each design is the binary nsymm value with bs = 4, followed by nh lines which are the x,y,r triplets, with bx, y = 10, br = 3. Mutation rates used were ps = (10bs) − 1, px, y = (4bx, y) − 1, pr = (4br) − 1. In (a), only the symmetry of this design has been mutated from 0010 (nsymm = 2) to 1011 (nsymm = 11). In (b) and (c) some modifications have been made in the positions of the holes, but not the genotypic sizes of the holes, that is, the last 3 bits for each hole have not been mutated, and changes in the sizes of the holes are a result of the embryogeny. (d), (e) and (f) illustrate some more complex design examples.

2.6 Measuring Fibre Fitness

The discussion in this chapter has so far explored the specific features of the GA to design microstructured optical fibres. The remaining aspect is the measure of fitness used to compare designs. Technologically relevant design objectives include confinement loss in the case of single-mode fibre, and bandwidth for the case of multimode fibres, for example. However, these design objectives are too computationally expensive for the exploration of the performance of the GA. Running the GA multiple times over thousands of generations using a full vector electromagnetic modelling algorithm would require hundreds of thousands of CPU hours. A simpler objective function, therefore, was devised in order to explore the properties of the GA, and this is explored in the remainder of this chapter.

2.6.1 Using an Approximate Fitness Function for Bandwidth.

In solid core multimode fibres, bandwidth is optimal when a parabolic graded index profile is used. Graded index profiles $n(r)$ are typically defined in the literature (Ishigure *et al* 1996) as

$$n(r) = n\infty \sqrt{1 - 2\Delta \left(\frac{r}{z}\right)^{\theta}}, 0 \leq r \leq a \tag{3}$$

where a is the core radius, Δ is the refractive index contrast and n_{co} is the refractive index of the core. Values of $g = 2$ are typically cited (Ishigure *et al* 1996, Ishigure *et al* 2003) as optimal with respect to maximising bandwidth.

In the move to microstructured multimode fibres, the idea of a graded index can be retained since a large array of holes can be designed to approximate a circularly symmetric parabolic profile $\bar{n}(r)$. So, rather than using complex electromagnetic mode solvers to evaluate the bandwidth, this simple approximation is used. The aim is to design microstructured fibres where the holes form, on average, an optimal parabolic profile $n_{opt}(r)$ (Equation 3). A core index of PMMA at $\lambda = 650$ nm is used ($n_{co} = 1.498$), and a cladding index of air with $n_{cl} = 1.0$.

The average index profile is evaluated by dividing the fibre into a series of N equally spaced concentric rings with inner and outer radii r_i and $r_i + 1$ respectively. The average refractive index of each ring then corresponds to

$$\bar{n}_i(r) - 2\pi \int_{r_i}^{r_{i+1}} n(r)rdr \tag{4}$$

The average difference between the averaged profile and optimal profile - the objective function, is defined as

$$f = \frac{1}{N} \sum_{i=1}^{N} |n_i - \bar{n}_i| \tag{5}$$

which is to be minimised. Five examples of random microstructured fibre designs along with their average refractive index profiles using $N = 100$ are shown in Figure 5-14. Each generation takes approximately 3 seconds to evaluate using a parallelised PThreads GA implementation on a dual-CPU Pentium 4 computer.

3.   EVOLVING MICROSTRUCTURED FIBRES: A SIMPLE TEST STUDY

An important property of a GA is the ability to find optimal designs with similar objective values, over multiple runs. In the presence of the stochastic nature of the starting population, the recombination, mutation and tournament selection operators, the GA must converge to similar solutions. This helps ensure that the optimal solutions found do correspond to globally optimal or near optimal solutions.

To complete the GA implementation, the fast fitness function discussed in the previous section is used which approximates bandwidth, and is also technologically relevant. Examining the properties of the GA over time can also yield insights into the behaviour of the objective values, and other properties specific to the representation used, such as the change in symmetry and complexity of designs over time.

3.1   Parameter Selection

There is a large body of research which explores the appropriate selection of parameters for GAs (Eiben & Smith 2003, pp. 129–151), but tuning them for a particular fitness function is not necessarily optimal for other fitness functions. Further, given the computational complexity of the fitness functions explored in this work, this exploratory approach is impractical. Given that this GA will be used to explore 3 different problems (including work reported in (Manos 2006)), parameter values are selected using results from simple preliminary experiments and values quoted from literature for binary genotypes. The parameters used in this test case are summarised in Table 5-2.

Population sizes of $N_p = 50$, $N_c = 50$ are used. A total population size of 100 is typical in GA applications, as is the use of a larger parent pool is to maintain a larger, more diverse parent population. The population is initialised such that each individual can have up to 15 holes defined in the genotype. The mutation rates are chosen according to the binary length of each binary segment. In order to create a diverse range of structures, a single bit of n_{symm} is mutated per individual on average. The same argument follows for the x_i, y_i and r_i representations, but lower mutation rates per bit are used to help preserve existing hole positions and sizes. On average, one hole per radial segment is deleted and one hole added per individual every generation. In order to prevent bias of outer regions of the genotypes being recombined, every individual has two randomly selected binary hole segments swapped per generation. The recombination operator does not require any parameters.

3.2   Results

The GA was run 5 times using randomly generated starting populations and the parameters outlined in Table 5-2. The evolution towards minimising the objective function (Equation 4) for all runs is shown in Figure 5-11. The elitist selection scheme is evident by the stepping nature of the plot - optimal designs are only replaced when better ones are found.

The remaining discussion focuses on a single run from Figure 5-11, but is equally valid for all runs.

Table 5-2.   Parameter settings for testing of the micro-structured fibre genetic algorithm.

Population and overall run details	
Population size	$N_p = 50$, $N_c = 50$
Maximum generations	10,000
Genotype and embryogeny related values	
Symmetry n_{symm}	$b_s = 5$, $n_{min\_symm} = 2$, $n_{max\_symm} = 31$
Position x_i, y_i	$b_{x,y} = 10$ (7 whole, 3 fractional) $\min\ \{x_i,\ y_i\} = 0.0\,\mu m$ $\max\ \{x_i,\ y_i\} = 127.875\,\mu m$
Hole sizes r_i	$b_r = 3$, 8 different sizes (Table 5-1)
Fibre reduction factor	340
Wall thickness constraint	$w_h = 0.2\,mm$
Maximum initial holes	15
Mutation rates	
Symmetry	$p_s = 1/b_s = 0.2$
Position	$p_{x,y} = 1/(2b_{x,y}) = 0.05$
Size	$p_r = 1/4b_r = 0.083$
Hole addition	$p_a = 0.01$
Hole deletion	

Figure 5-12 summarises the behaviour of the objective value over the parent and child populations. As the GA progresses, the parent population is continuously refined. The average objective value improves, along with a convergence of the maximum and minimum values, and the standard deviation is indicative of convergence towards a population with similar objective values. In contrast to this, the child population maintains quite high diversity in the objective values produced. The GA maintains a diverse child population through recombination and mutation, while continuously improving the (elitist) parent population.

Figures 5-14, 5-15, 5-16 and 5-17 show the top 5 designs at generations 0, 100, 1000 and 10000 respectively. The GA has generally favoured high symmetry designs, and by generation 10000 it has discovered that designs with

$n_{symm} \approx 18$ and approximately 130 holes in total is most optimal. The top five designs (Figure 5-17) are slight variations of this theme.

Also of interest is the evolution of various features of the embryogeny developed for this GA, summarised for the parent population in Figure 5-13. Up to generation 100, the GA can be seen to be favouring designs with high symmetries, indicated by the increase in the average n_{symm}. Beyond this point, the parents converge to similar symmetry values.

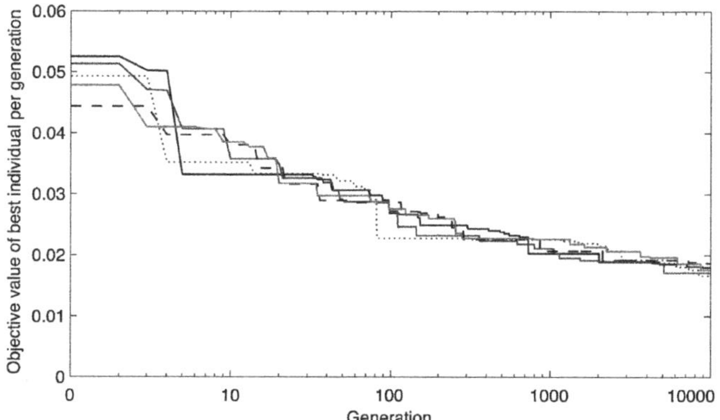

Figure 5-11. Evolution of the objective value (Equation 4) for five different runs of the GA using random starting populations, over 10,000 generations. The objective value shown is that of the best individual per generation, the generation axis is a $\log_{10}$ scale.

The evolution of the length of the binary genotype shows an initial favouring of longer genotypes, which then reduces from generations 10 to 100, and increases again. A similar pattern is seen for the total number of holes encoded. This corresponds to simple designs consisting of a few rings of holes initially being found optimal. Over time, through recombination and mutation, adjustments are made to these designs by adding more holes, making them more complex. This corresponds to the change in phenotypes seen in Figures 5-15, 5-16 and 5-17.

The number of expressed holes over time refers to encoded holes which have a non-zero radius, and corresponding to this is the ratio of unexpressed to expressed holes. As the GA progresses, this value decreases, suggesting that optimal designs make more efficient use of the genotype, such that almost all encoded holes map to a phenotypic hole. This is also a feature of the best designs, which generally have no unexpressed holes. Finally, the distinct holes sizes refer to the number of different holes sizes expressed in the phenotype (Table 5-1). The increase in fitness of the parent population corresponds to a more diverse range of holes being used in the optimal microstructure fibre designs.

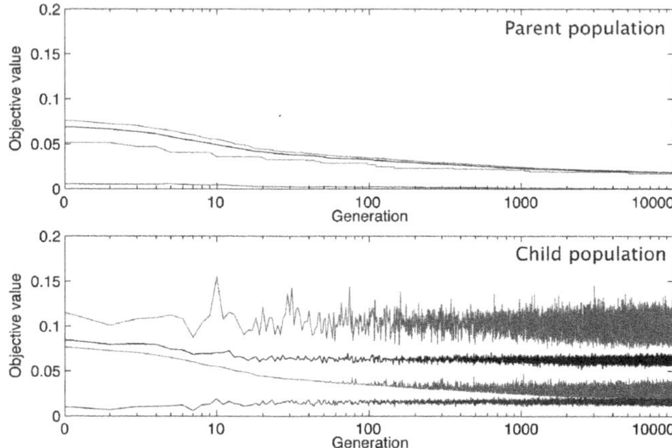

Figure 5-12.　　Typical behaviour of the objective value over time. The top graph shows the parent population. Shown is the maximum (worse) objective (top red line), average parent objective value (black), minimum (best) objective (bottom red line), and the standard deviation (blue). Convergence to a much fitter parent population over time can be seen. The bottom graph shows the same details but for the child population only.

With respect to the child population, diversity in the symmetry is high for every generation, where designs are breed with symmetries $n_{symm} = 2, \ldots, 31$. Features, such as the average genotype length, gradually increasing over time corresponds to the features evident for the parent population. This is expected since the children are offspring of the parents, but the maximum and minimum values are far more diverse, ranging from a single encoded hole to 10 encoded hole designs in each generation.

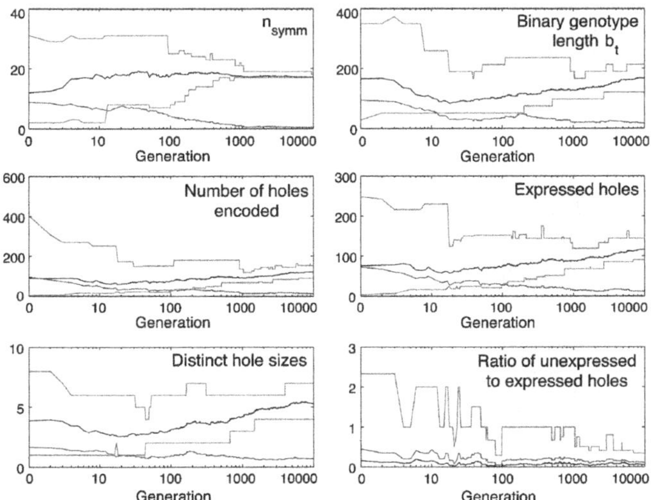

Figure 5-13. Behaviour of various parameters associated with the parent population over time. In each plot the following values are shown: the maximum (top red line), average value (black), minimum (bottom red line), and the standard deviation (blue).

The advantages of a variable length genotype are clear - the GA has evolved designs from simple microstructured fibres to fitter designs with more holes and more complicated refractive index profiles.

Figure 5-18 shows two photographs of real optical fibres, made using evolved designs.

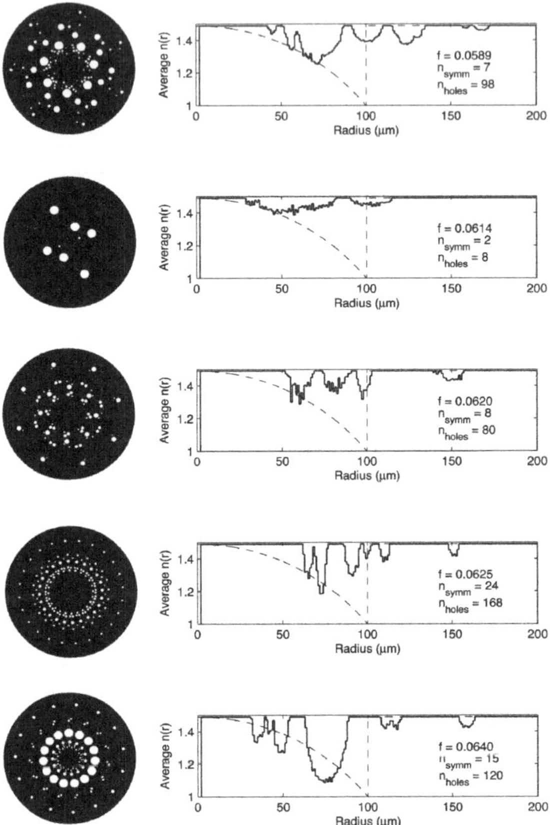

Figure 5-14. Best 5 individuals of the randomly generated initial population (Generation 0). The fibre cross sections are shown on the left. The resulting averaged profiles are shown on the right, with the target profile (dashed line) along with fibres averaged profile (solid line). The corresponding fitness values, symmetries and number of expressed holes are shown.

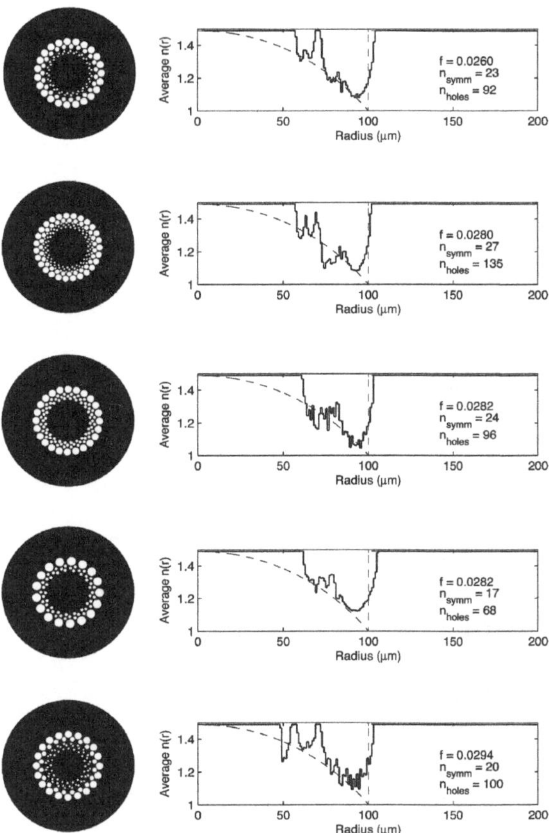

Figure 5-15. Best 5 individuals of Generation 100. The best designs have started evolving to high symmetries, causing the designs to best suit the outer edges of the parabolic index profile.

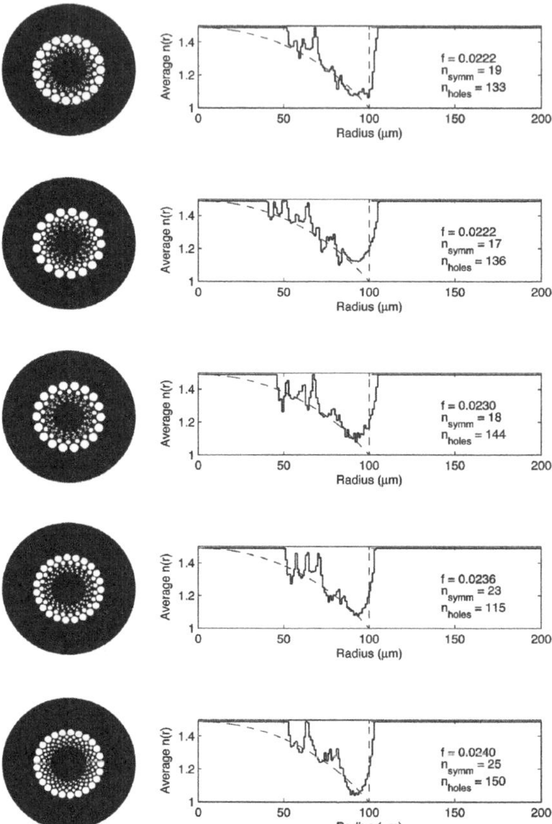

Figure 5-16.   Best 5 designs of Generation 1000. Optimal designs have approximated the parabolic profile closely, where many small holes have been added in the central core region to further refine the average profile.

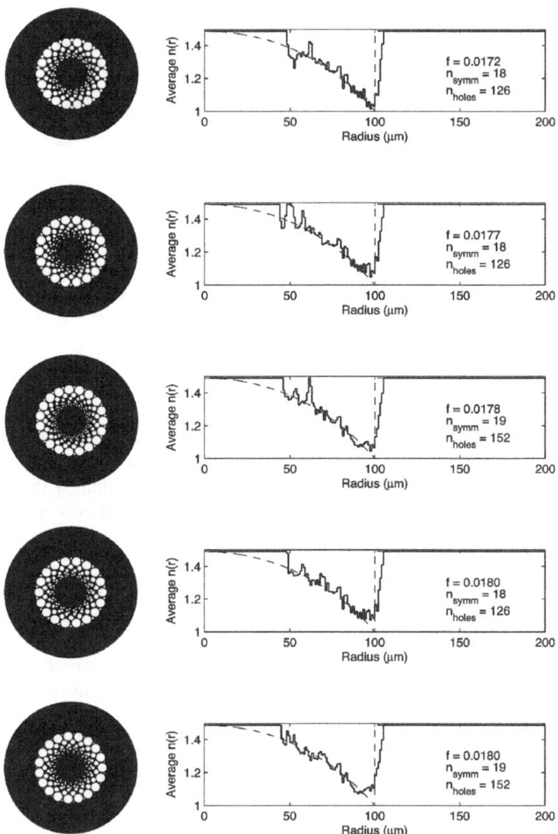

Figure 5-17.   Best 5 designs of Generation 10000. Designs have settled into an optimal symmetry of n-symm = 18, 19, where the different variations of the best designs mainly differ in the details of the central air holes.

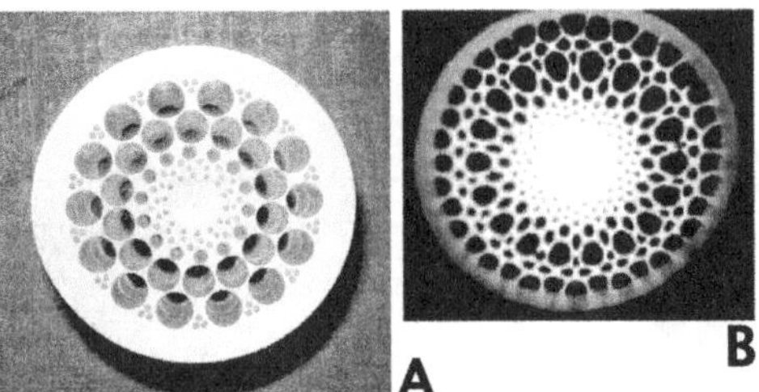

Figure 5-18.   A photograph of the drilled perform of an evolved fibre design. The perform is 70 mm in diameter, made from PMMA (polymethylmethacrylate) polymer and contains 126 holes with n-symm = 14. This is then drawn down to an optical fibre, reduced in diameter by a factor of approximately 340. Photograph B shows another evolved design which was found to be easier to draw to fibre, which had accurately maintained its structure during the draw. This fibre is 220 μm (1/5 mm) in diameter, which guides light down the central core region.

4.   DISCUSSION

Optimisation and design refer to two different processes. Variations of a fixed-length genotype that use a direct mapping result in a simple optimisation, where different candidate designs are qualitatively similar. In contrast, variations of the variable length microstructured fibre genotype presented in this chapter resulted in the process of design, where both the features and the complexity of design can be automatically evolved over time. Microstructured optical fibres have traditionally focused on a fixed design theme – a hexagonal array of holes. By using a more expressive representation, many more types of microstructured fibres can be explored, resulting in a new design, rather than just optimising the details of a hexagonal array.

The GA has also been used to design single-mode microstructured optical fibres, resulting in the evolution of designs which use different mechanism to achieve single-mode guidance. Instead of using an approximate fitness function, electromagnetic modelling software is used to determine the characteristics of modes within fibre designs, such as their confinement loss. A multi-objective approach was used, and designs were found to require fewer holes than previous designs to achieve equal or better performance. Thus, the variable complexity was advantageous in finding simpler microstructured designs. The multi-objective approach is particularly powerful since large changes in design can be examined with respect to changes in the design objectives (electromagnetic properties). Traditional single-mode microstructured fibres rely on hexagonal arrays of holes, which form a low-contrast cladding, emulating traditional low-contrast step-index fibres. A large number of rings of holes are required in this case, and without these prior assumptions, single-mode designs, with fewer holes, were evolved. Each individual fitness evaluation can take up to 20 minutes or more, however, the use of high performance computing clusters delivers optimal designs within a few days, and GA-design fibres of interest are currently being patented.

The analysis of optimal designs generated in both the single-mode design problem and the graded-index problem has revealed information about the origin of holes, which becomes evident in two aspects of the microstructured fibre representation. Firstly, holes which are phenotypically expressed, but do not contribute to the objectives (for example, by being located far away from the core), and secondly, dormant holes which are not expressed at all. The former primarily relates to selection, since particular phenotypic features may not influence the objectives as strongly as others. The latter relates to the behaviour of the representation. The existence of dormant holes could play a vital role in evolution, appearing later on as phenotypic features that improve the fitness of a design. Even though evolution may proceed for some generations without improvement in fitness, the underlying gene pool in the population can undergo re-organisation of neutral genotypes (those which do not increase the fitness), resulting in improvements later on (so called neutral networks).

Understanding all these processes will lead to the answer of whether designs, such as the single-mode microstructured fibres, are globally optimal. It also shows that although the representation is much more expressive than previous examples, it may still have a predisposition, where particular types of designs are easier to evolve than others. However, the expressiveness is most noteworthy when comparing the two problems considered. Graded-index and single-mode microstructured fibre design all resulted in distinctly different types of fibres, using the same underlying representation, and only differing in the fitness function.

5.   CONCLUSION

This work has successfully demonstrated the application of variable-complexity, evolutionary design approaches to photonic design. The inclusion of real-world constraints within the embryogeny aids in the manufacture of designs, resulting in the physical construction and experimental characterisation of both single-mode and high-bandwidth multi-mode microstructured fibres, where some GA-designed fibres are currently being patented. With further improvements in the expressiveness of the representation, the improvement of photonic devices using these design approaches will no longer be limited by human imagination.

Acknowledgements

The authors would like to acknowledge the contributions of and helpful discussions with Maryanne Large (Optical Fibre Technology Centre, University of Sydney, Australia) and Leon Poladian (School of Mathematics and Statistics, University of Sydney, Australia).

References

Bentley, P. J. and Kumar, S. (1999) Three ways to grow designs: A comparison of embryogenies for an evolutionary design problem. In Proceedings of the Genetic and Evolutionary Computation Conference (GECCO 1999), 13–17 July 1999, Orlando, Florida, USA, pages 35–43.

Birks, T. A., Knight, J. C. and Russell, P St J. (1997) Endlessly single-mode photonic crystal fiber. Optics Letters, 22(13):961–963, July 1997.

Deb, K. Agrawal, S, Pratap, A and Meyarivan, T. (2000) A fast elitist non-dominated sorting genetic algorithm for multi-objective optimization: NSGA-II. Proceedings of the Parallel Problem Solving from Nature VI (PPSN-VI), pp. 849–858.

Eiben, A.E. and Smith, J.E. (2003) Introduction to Evolutionary Computing. Springer-Verlag.

Goldberg, D. E. (1989) Genetic Algorithms in Search, Optimization & Machine Learning. Addison-Wesley.

Ishigure, T., Nihei, E. and Koike, Y. (1996). Optimum refractive-index profile of the graded index polymer optical fiber, toward gigabit data links, Applied Optics 35(12): 2048–2053.

Ishigure, T., Makino, K., Tanaka, S. and Koike, Y. (2003). High-bandwidth graded-index polymer optical fiber enabling power penalty-free gigabit data transmission, Journal of Lightwave Technol. 21, 2923.

Issa, N. and Poladian, L. (2003) Vector wave expansion method for leaky modes of microstructured optical fibres. Journal of Lightwave Technology, 22(4):1005–1012.

Knight, J. C., Birks, T. A., Russell, P. St. J. and Atkin, D. M. (1996) All-silica single-mode optical fibre with photonic crystal cladding. Optics Letters, 21(19):1547–1459, October 1996.

Kominsky, D., Pickrell, G. and Stolen, R. (2003) Generation of random-hole optical fibers. Optics Letters 28(16), August 2003.

Large, M. C. J., Ponrathnam, S., Argyros, A., Bassett, I., Punjari, N. S., Cox, F., Barton, G. W. and van~Eijkelenborg, M. A. (2006) Microstructured polymer optical fibres: New Opportunities and Challenges. Journal of Molecular Crystals and Liquid Crystals 446:219–231.

Manos, S. and Poladian, L. (2002) Optical Fibre Design with Evolutionary Strategies: Computational implementation and results. 4th Asia-Pacific Conference on Simulated Evolution and Learning, November 18–22, Singapore.

Manos, S. (2006) Evolving Fibres Designing Fibre Bragg Gatings and Microstructured Optical Fibres using Genetic Algorithms. PhD Thesis, Optical Fibre Technology Centre and School of Physics, University of Sydney.

Monro, T. M., Bennett, P. J., Broderick, N. G. R. and Richardson, D. J., (2000) Holey fibers with random cladding distributions. Optics Letters 25(4): 206–208.

Poladian, L., Issa, N.A. and Monro, T. (2002) Fourier decomposition algorithm for leaky modes of fibres with arbitrary geometry. Optics Express 10; 449–454.

Poletti, F., Finazzi, V., Monro, T M., Broderick, Tse, V. and Richardson, D. J. (2005) Inverse Design and Fabrication tolerences of ultra-flattened dispersion holey fibers. Optics Express 13(10):3728–3736, May 2005.

Ranka, J. K., Windeler, R. S. and Stentz, A. J. (2000) Visible continuum generation in air-silica microstructure optical fibres with anomalous dispersion at 800 nm. Optics Letters 25(1):25–27, 2000.

Renversez, G., Kuhlmey B. and McPhedran, R. (2003) Dispersion management with microstructured optical fibres: ultraflattened chromatic dispersion with low losses. Optics Letters 28(12):989–991, June 2003.

Stanley, K. O. and Miikkulainen, R. (2003) A taxonomy for artificial embryogeny. Artif. Life, 9(2):93–130.

van Eijkelenborg, M., Large, M., Argyros, A., Zagari, J., Manos, S., Issa, N. A., Bassett, I. M., Fleming, S. C., McPhedran, R. C., de Sterke, C. M. and Nicorovici, N. A. P. (2001) Microstructured polymer optical fibre. Optics Express, 9(7):319–327, September 2001.

van Eijkelenborg, M. A., Poladian, L. and Zagari. J. (2001) Optimising holey fibre characteristics. In CLEO/Pacific Rim: Proceedings of the 4th Pacific Rim Conference on Lasers and Electro-Optics, volume 1, pages 436–437.

van Eijkelenborg, M. A., Argyros, A., Bachmann, A., Barton, G., Large, M. C.J., Henry, G., Issa, N. A., Klein, K. F., Poisel, H. Pok, W., Poladian, L., Manos, S. and Zagari, J. (2004) Bandwidth and loss measurements of graded-index microstructured polymer optical fibre. Electronics Letters 40(10): 592–593, May 2004.

van Eijkelenborg, M., Argyros, A., Barton, G., Bassett, I., Fellew, M., Henry, G., Issa, N., Large, M., Manos, S., Padden, W., Poladian, L. and Zagari, J. (2003) Recent progress in microstructured polymer optical fibre fabrication and characterization. Optical Fiber Technology 9; 199–209.

Xue, S. C., Large, M. C. J., Barton, G. W., Tanner, R. I., Poladian, L. and Lwin, R. (2006) Role of Material Properties and Drawing Conditions in the Fabrication of Microstructured Optical Fibres. Journal of Lightwave Technolgy 24(2): 853–860.

Yu, T. and Bentley, P. J. (1998) Methods to evolve legal phenotypes. In PPSN V: Proceedings of the 5th International Conference on Parallel Problem Solving from Nature, pages 280–291, London, UK. Springer-Verlag.

Chapter 6

MAKING INTERACTIVE EVOLUTIONARY GRAPHIC DESIGN PRACTICAL

A Case Study of Evolving Tiles

Carl Anderson[1], Daphna Buchsbaum[1], Jeff Potter[1], and Eric Bonabeau[1]

[1]*Icosystem Corporation, 10 Fawcett Street, Cambridge, MA 02138, USA*

Abstract
This chapter describes interactive evolutionary design, a powerful technique where one marries the exploratory capabilities of evolutionary computation with the esthetic skills and sensibility of the human as selective agent. Interactive evolutionary design has the potential to be an enormously useful tool to graphic designers. However, in order for it to become commonly used, a number of barriers to use must be overcome, including making it both simpler to understand (e.g., more intuitive genotype-phenotype mappings) and with greater user-control (e.g., allow the user to lock "perfect" elements of a design). In this study we explore these ideas—how can one make interactive evolutionary design more appealing, useful and practical?—by developing a design tool "evoDesign" that employs a genetic algorithm to evolve designs. As an illustrative test case, evoDesign is used to evolve tiles used for walls or floors or as a repeating unit for fabrics and wallpapers.

Keywords: Interactive evolution; graphic design; tiles; evolutionary art.

1.   INTRODUCTION

Evolutionary art, with human esthetic choice as the driving selective agent, has been around for many years and has been used to produce some stunning visual images (McCormack 1993; Rooke 2005; Sims 1991; Todd and Latham 1992; Whitelaw 1999, 2004). However, these images, which are often very organic and fractal-laden, are not necessarily appropriate for many practical applications; for example, they may not be something one would want to tile a bathroom with. In most cases, these artistic explorations are conducted by professional computer scientists or artists, people who are striving to create "foreground art" and who need not be concerned by mass appeal and salability, nor with the usability of their evolutionary tool.

C. Anderson et al.: *Making Interactive Evolutionary Graphic Design Practical*, Studies in Computational Intelligence (SCI) **88**, 125–141 (2008)
www.springerlink.com

However, interactive evolutionary design (Takagi 1998, 2001) has the potential to be an enormously useful tool to graphic designers and individuals designing clothing and other goods for mass markets, *if* it can be made more practical and palatable to those practitioners, many of whom do not have a quantitative background. In this study, we explore some of the issues of making 2-D graphic design through interactive evolution more practical and appealing in the real world, focusing on a tool for evolving floor, wall and fabric tiles as an illustrative test case.

In order for interactive evolution to become a commonly used graphic design tool, a number of barriers-to-use presented by existing systems must be overcome. One of the critical problems of work to date, at least in terms of interactive evolution as a *practical* tool for graphic designers, is that most evolutionary art is based upon genetic programming. This can be a very hard concept to grasp and to explain to non-expert end-users; for many people, genetic programming is very mathematical and very abstract. A genetic algorithm (GA) on the other hand is far simpler and *encapsulates* the same set of actions that a graphic designer performs when creating and modifying a design: adjusting the position or rotation of an element, altering the hue, transparency or color of a component and so on. For this reason, we use a simple GA rather than genetic programming.

Another problem with work to date is that it is often hard to control individual elements of a graphic independently. When using genetic programming, the picture is usually generated from a single equation (e.g. a LISP expression) in which all the elements are intertwined; it is often hard to fix one esthetically pleasing element while allowing other elements to evolve (this is also true of other multi-equation systems such as L-systems)—in short, genetic programming is brittle. This has two important consequences. First, a graphic designer may feel a loss of control. The goal of interactive evolutionary design is to marry the exploratory skills of evolutionary computation with the esthetic skills of the human as selective agent. A user who sees a pleasing element but who is unable to retain it in the design is likely to get frustrated with the system very quickly. Second, and related, the designer may feel threatened by this technology as a tool that is taking away their job. The goal should be to develop a tool that is complementary, perhaps supplementary, to the designer. The designer should be in control, either using the tool to generate some initial ideas or using interactive evolution to explore the design space around an already existing idea. The designer must be the most powerful driver of the system through the design space.

In this study, we attempt to put ourselves in the mind of a skilled graphic designer, one interested in exploring a particular design space, perhaps using the tool to generate initial ideas within some user-defined boundaries. Alternatively, they might be starting with a pre-existing idea, some graphical "digital amber" (Rooke 2005; Whitelaw 2004), which they would like to perfect. They want to be able to lock "perfect" elements as they evolve (Takagi 1995), or even edit them directly. And, they would like to understand the basic principles of the system. In other words, we develop a system that is intuitive, useful and practical, and (in hindsight) adopt a number of the "ten steps to make a perfect creative evolutionary design system" (Bentley and O'Reilly 2001) resulting in a system that happens to be much simpler to understand and intuitive to use.

2. GENETIC ALGORITHM

2.1 Phenotype

In this design tool, a phenotype consists of a panel, a "tile" upon which is drawn a set of simple graphical elements: colored horizontal and vertical stripes, circles, squares and rectangles. Fours such tiles are shown in Figure 6-1. Each tile is toroidal or wrapped so that parts of elements that overlap one margin appear at the opposite margin. This guarantees that all designs that evolve constitute a repeatable unit (i.e., are tile-able in the strict mathematical sense) that is suitable for fabrics, wallpapers and so on (but see later discussion). The evoDesign tool contains a feature that allows one to see a chosen panel tiled across the whole screen at a user-controlled density (see Figure 6-2).

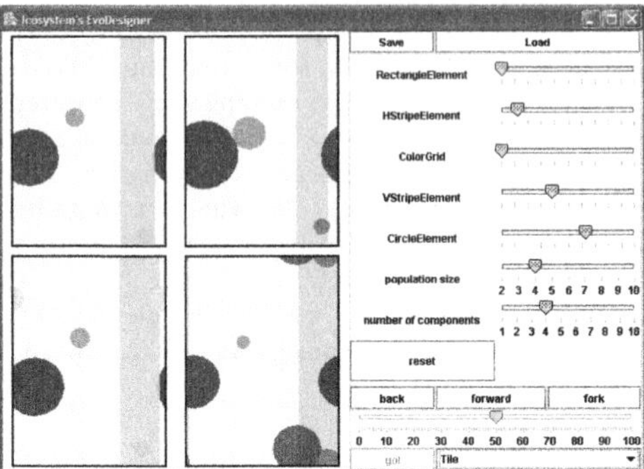

Figure 6-1.   The evoDesign user interface. The four panels on the left constitute a population of tiles being evolved. On the right, sliders determine the relative proportion of the different graphic element types, the next two sliders, the number of panels and the number of graphic elements within each panel. "Back" and "forward" allow the user to revisit previous generations.

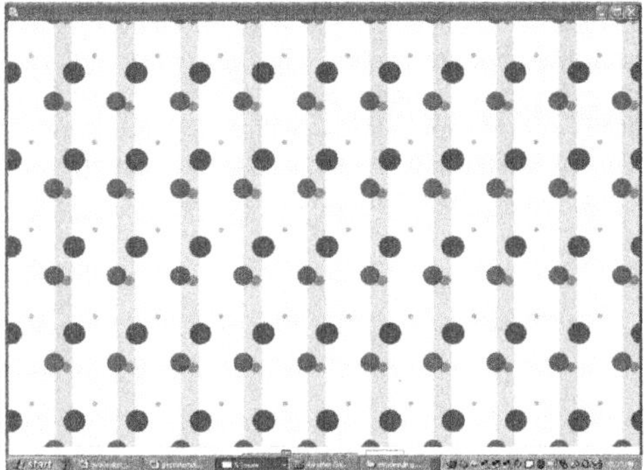

Figure 6-2.   A tiling of the bottom right tile of figure 6-1.

2.2   Genotype

A genotype i, G_i, consists of a vector of graphic elements ("chromosomes")—an individual circle, square etc.—each of which has a number of evolvable attributes ("genes"). Every graphic element has an evolvable four-gene color consisting of a red, green, blue (RGB) and α-(transparency) value, all in the range of $\{0, 255\}$. The evoDesign tool is used to evolve the number, size, shape, location, color and transparency of the graphical elements within a set of tiles.

2.2.1   Circle.   A circle contains three other evolvable attributes: an (x, y) coordinate for the top left of the circle's bounding box—the panels are normalized so that all x and y values are $(0,1)$—and a bounded diameter $d \in (0.05, 0.5)$. In short, a circle's genotype is the vector $(R, G, B, \alpha, x, y, d)$.

2.2.2   Rectangles and Stripes.   A rectangle contains four other evolvable attributes: an (x, y) coordinate for the rectangle's upper-left corner, a width (w) and a height (h), both with a lower bound of 0.05. In short, a rectangle's genotype is the vector $(R, G, B, \alpha, x, y, w, h)$.

A vertical stripe is simply a special case of a rectangle where $y = 0$ and $h = 1$, both of which are non-evolvable. Similarly, a horizontal stripe is a special case of a rectangle where $x = 0$ and $w = 1$, again, both are non-evolvable. The maximum width of a stripe is 0.3.

2.2.3   Color Grid.   A color grid consists of a regular grid, usually 4×4, of colored rectangles. The color grid's only evolvable attributes are the color and transparency of its component rectangles.

The color, size and position of all graphical elements (except those of a color grid) in a tile may be "frozen," i.e. made non-evolvable, from the user interface.

A particular gene may be frozen per element (e.g., the color of one circle) or throughout the entire tile (e.g., all elements have their color frozen, while size and position continue to evolve).

2.2.4 Drawing. The elements of G_i are drawn consecutively on tile i with later elements potentially overlapping or obscuring earlier elements, where the color grid constitutes one element. Hence, a final implicit attribute of a graphical element within G_i is its position or index which corresponds to a graphical "level."

Except for frozen (x, y) coordinates, all attributes of graphical elements are independent. Hence, initialization of a tile simply consists of generating a given number of graphical elements, each with a random (bounded) value.

2.3 Mutation Operator

A genotype is mutated by randomizing the order of its graphic elements and then mutating each element. In addition, with a certain probability (here, $^1/_3$) a randomly chosen element is removed from the genotype and with a certain probability (here, $^1/_3$) a new graphic element is created and added to the genotype.

Each non-frozen gene that is mutated is done so uniformly across some half-range r around its current value and then bounded as necessary. That is, a gene g with current value $g(t)$ and bounds g_{low} and g_{high} takes on a new value,

$$g(t + 1) = \max(g_{low},\ \min(U(g - r,\ g + r),\ g_{high})). \tag{1}$$

The following default values are used: $r_x = r_y = 0.05$; $r_R = r_G = r_B = 5$; $r_\alpha = 60$, $r_d = 0.1$, and $r_w = r_h = 0.05$.

2.4 Crossover Operator

In addition to mutation, we also use a crossover operator (Bentley 1999). Suppose that one parent genotype G_i contains n_i graphic elements while the other parent genotype G_j contains n_j elements and that $n_i \leq n_j$. Their offspring

G_k will have $n = U\{n_i, n_j\}$ genes, i.e. a length between that (inclusive of the bounds) of the two parents. The first $U(0, n_i)$ of which are randomly chosen (without replacement) from G_i with the remainder randomly chosen from G_j. In brief, an offspring is created of similar length to its parents, with some random proportion of its components coming from one parent and the remainder from the other. Figure 6-3 gives an example of the crossover operation.

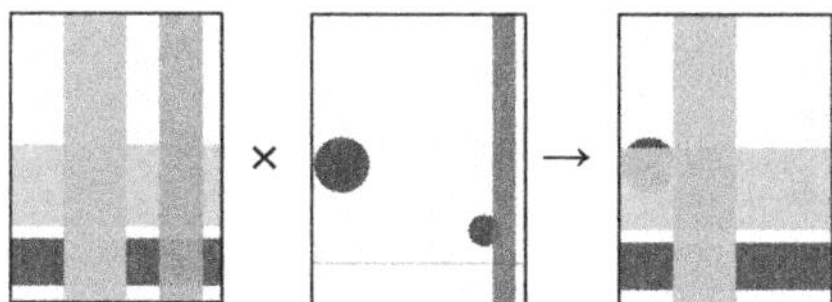

Figure 6-3. Crossover between two parents.

3. INTERACTIVE EVOLUTION

The evoDesign tool contains a population of 2 to 10 genotype tiles (Figure 6-1 contains a population of size 4). The user can evolve the population in a variety of ways. First, she can use mutation only. In this case, the user selects a tile which is then passed unaltered into the next generation with mutant offspring derived from this parent making up the remainder of the population. Alternatively, the user can evolve the population by crossover. Here, the user selects two or more parents and the next generation is entirely filled by their offspring (as described above). When more than two parents are selected for crossover, each offspring is created by crossing two of those parents (chosen randomly). Finally, the user can choose mutation and crossover in which each offspring is either created by mutation or by crossover, the probability of which is determined by the lowest slider in Figure 6-1. At any point, the user may select a tile for editing in which they can freeze the shape, size and color of any (or all) of the tile's components (see Figure 6-4).

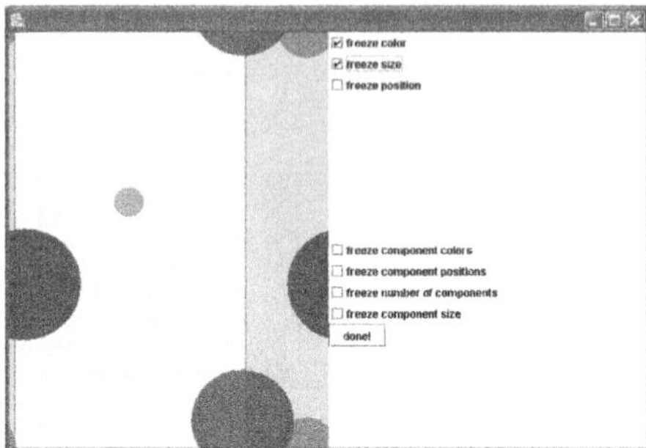

Figure 6-4.   The bottom right tile in figure 1 is edited: here the vertical stripe is selected and its color and size frozen.

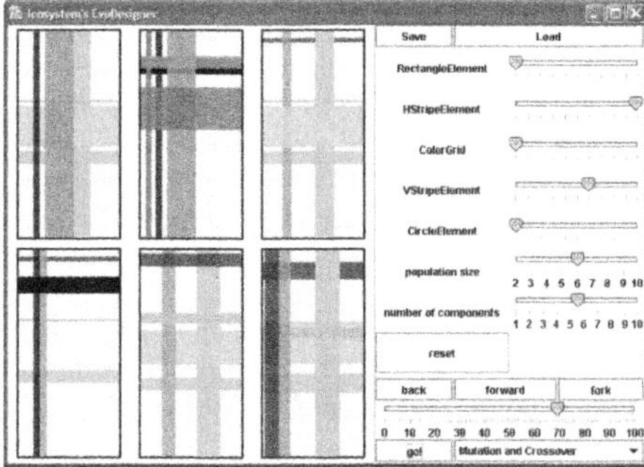

Figure 6-5.   Example plaid tiles with a relatively high proportion of vertical and horizontal stripes.

3.1.1   Initialization.   Initialization occurs by either loading a population of tiles from a file, or by randomizing the population. A slider determines the number of graphical components within a tile and other sliders determine the relative proportion of the different graphic component types. Thus, one can easily increase the absolute number and relative proportion of vertical and horizontal stripes to explore plaid patterns (see Figures 6-5 and 6-7) or circles for more retro designs (see Figures 6-6 and 6-7).

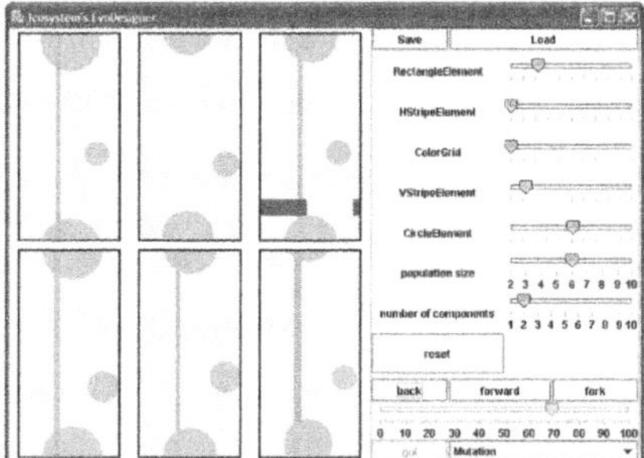

Figure 6-6.   A retro pattern with a low number of graphical components but high proportion of circles.

3.1.2   Exploring Design Space.   In interactive evolution the user usually does not know in advance quite what they are after; it's often a case of "I'll know it when I see it." However, as in any exploratory activity, one can be led down a blind, unfruitful alley. For this reason, the evoDesign tool includes a back button that allows one to revisit previous generations. Thus, one can explore one avenue, decide that this is not a desirable portion of the design space and return to an ancestor population of designs to explore other avenues using the mutation and crossover operators as before. Additionally, the save/load feature can be used with a similar effect and there is also a fork button that allows one to spawn off a new instance of the tool with the same current design population. This ability to revisit any point in the current run's "design tree" is particularly useful and important; without it, it can be surprisingly frustrating when a design is not evolving satisfactorily. Despite the seemingly limited number of graphical components, one can evolve a surprising variety of pleasing design (something one *would* want to wear to work or tile a bathroom with), and ones that represent or evoke different styles and decades. A few "samples" are shown in Figure 6-7.

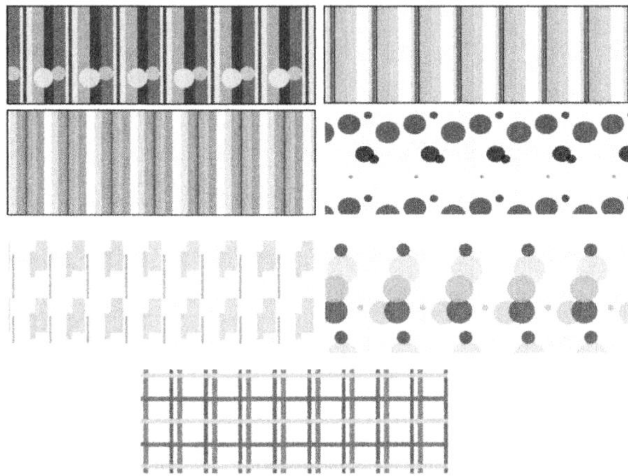

Figure 6-7. Some evolved wallpaper or fabric "samples".

O'Reilly and Testa (unpublished), in the context of interactive evolution, suggest that "architects are interested in families of designs rather than finding one absolute best." Similarly, other designers are interested in designing a set of pieces which together form a coherent collection. Each new piece has to conform to that collection's or designer's signature style but it is hard to define precisely what that is. evoDesign can be used to such effect. Figure 6-8 shows a target design, Paul Smith stripes, and a population from evoDesign that has a similar Paul smith look and feel.

4.   DISCUSSIONS

Interactive evolution (using GAs) has been used in a number of recent 2-D graphic-related applications, including evolving color schemes (Kelly 1999), photofit composite images (Frowd et al. 2004), anthropomorphic symbol designs such as a warning sign (Dorris et al. 2003), and product branding and marketing (see affinova.com). However, the full potential of interactive evolutionary creative design has yet to be realized.

What we set out to do was to create a tool that was intuitive, useful and practical. The result happened to be a system that is significantly simpler to understand and easier to use. The target users of our case study were design experts because they would not only use such a tool more frequently than the general populace but because they also have the most to give: the system gets the user's expert esthetic and design expertise to judge "fitness" while the user gets the ability to explore around a design theme, within constraints (see Figure 6-8).

This simplicity, however, has profound implications not just for selling the idea to the experts—here, non-technical design experts—but also for ultimately making interactive evolution a commonly used technology in the general populace. There is almost no better way to promote a technology than to simplify it for the masses and give it away for (almost) free. Thus, Apple Computer's bundling of GarageBand in their iLife package has essentially commoditized music editing and recording. This has yet to be achieved with interactive creative design because previous systems are too application specific (and it is difficult to demonstrate the broader implications of the technology) and/or too complex for the majority of general users (non-computer scientists) to understand. While the general public doesn't need to understand all of the gears "under the hood" to be able to use the tool, it is helpful to provide them with the general information of the implications of their actions within the tool. In the case of evoDesign tool, the key is its ability to do genotype-phenotype mapping. In genetic art, most design tools are based on genetic programming, where the genotype of compact, abstract expression bears no resemblance to the final picture—the phenotype. In a simple genetic algorithm, such as that used in evoDesign, the mapping is much closer: what is being evolved and how those changes in the genotype affect the phenotype is much more closely related and easier to explain and to understand. Moreover, in order to be able to edit components of the design directly in the GUI and have those changes reflected in the genotype, it is crucial that the genotype mapping is 1:1 and completely reversible (O'Reilly and Testa, unpublished). The simpler the genotype, the more likely this is to happen.

Figure 6-8.   A target pattern: Paul Smith stripes and a population from the evoDesign tool.

4.1   An Ideal Interactive Evolution Tool

evoDesign is by no means perfect, but we believe is headed in the right direction, at least according to guidelines set out by Bentley & O'Reilly (2001) in their "ten steps to make a perfect creative evolutionary design system."

We have a specific domain 'in which it makes sense to use a computer for "creativity enhancement" ' (their rule 1). We have a "good reason for using a creative system at all" (rule 2). As discussed, a GA design tool can serve several sensible purposes: to generate new ideas or to explore the design space around a seed idea, and to do so in a manner that essentially mirrors the way that a designer might modify a design.

We believe that we have "appropriately balanced control of the design process between the tool's user and the tool itself" (rule 3)—the ability to freeze elements, to choose particular parents for crossover, to set the mutation rates and other parameters are all under the control of the users. However, a user-friendly editor in which a designer could create a specific initial seed tile, to interact with and modify *all* parameters of elements, and even to drag and drop elements, would be even better. The input and output format for the saved designs can be very easily tailored to systems that professional graphic designers use (rule 4). We hope we have demonstrated that evoDesign is "generative and creative" (rule 5) and "understandable" (rule 6). Because it is based on esthetic choice, and because of the ability to revisit designs in the previous generations, there is certainly "an easy and effective way of evaluating the quality of solutions and guiding the path of evolution" (rule 7). We are currently working on rule 8: "find people who are actually prepared to use the system". For completeness, we list the rest of their rules here: rule 9 is "get lots of money to pay R&D costs" and rule 10 is "start a company and make a billion."

4.2   Constrained Designs

In evoDesign, we included some simple design constraints such as the minimum and maximum width of stripes, the maximum radius of circles and so on. However, many designers are faced with more complex and subtle constraints; for instance, a client may have a particular defining color palette (say the red/white/blue/black of *Tommy Hilfiger*) or "feel and style"—e.g., *Versace* has a very definite style but is hard to define; again, this is a case of "I know it when I see it"—and the designer may only explore color-"equivalent" designs (*sensu* Feldman (2005)). Or, the design may need to evoke a particular era, say the bright colors of the 1960's or earthy tones of the 1970's. Such constraints can be incorporated into the GA. Kelly (2004) describes a GA-based application that he used to evolve color schemes. He explored several different types of constraints (Eckert et al. 1999), one of which is to average the color combinations in the scheme such that the colors in the design may evolve but under the constraint of the overall constraint of "greenishness".

Another important area of constraint, one that is not necessarily obvious but can be very important, is a design's manufacturability. Not all designs are alike: some designs are far easier and cheaper to produce than others, either because of the specific material required for a particular color whose color mixing gives the rate at which a product can be manufactured by a given machine, or there is a limit on the number of colors available for a machine such as a loom. However, some of these features can be subtle and not necessarily obvious in the final design—one may be able to make the design easier and cheaper to produce but have very little effect on the overall appearance and esthetic appeal, at least to the casual observer. Thus, it is possible to associate each design with a number of metrics, displayed in the user interface (Bandte and Malinchik 2004), and the user can then evolve designs that are both esthetically pleasing, satisfying esthetic design constraints (such as greenishness), and are also easy to manufacture.

4.3   Wallpapers

Our tiling feature (Figure 6-2) is a regular, rectangular lattice tiling. However, there are other ways in which one can tile a wall: just sixteen other ways in fact. Considering all the combinations of translation, reflection and rotation of a tile, mathematically speaking, there are only 17 types of wallpaper ("wallpaper groups"). For instance, one can place the tiles on a rectangular or hexagonal lattice, keeping all tiles in the same rotation (as in Figure 6-2) or rotate alternate tiles by $180°$ to form a checkerboard arrangement where white and black would represent tiles with different orientations and so on. What would these other wallpaper types mean for a tool such as evoDesign?

First, one would have to remove the toroidal nature of the tiles. That is, we wrapped the design around each tile so that they would be guaranteed to tile. However, while this works for this most basic wallpaper type, it does not hold true for all other wallpaper types. Let's consider the checkerboard arrangement. A tile that contains just one circle that overlaps the upper left margin of the tile would be positioned on the *lower right* margin of the adjacent tile; in short, they would not match up and there would no complete circles anywhere in the design. Thus, as a general rule across all wallpaper types, wrapping must be removed.

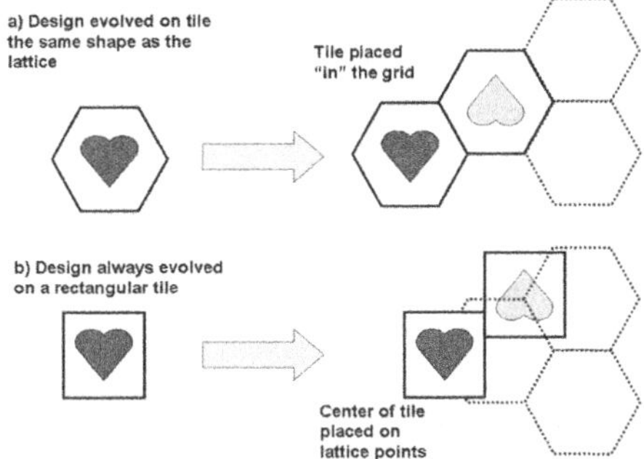

Figure 6-9. Two approaches to evolving tiles for the 17 different wallpaper types. a) evolving a tile of the same shape as the underlying lattice. b) evolving a rectangular tile and placing it on the lattice points.

Second, the tile shape could be altered. The underlying lattice in these 17 plane symmetry groups are parallelogramic and rhombic (the latter being a special case of the former thus they are essentially the same), rectangular and square (ibid.), and hexagonal. One approach would be to evolve designs on tiles of these different shapes; e.g. the panel would be a hexagram (see Figure 6-9a). Another approach would be to evolve designs on an unwrapped rectangle and then the center of that design would act as the pivot point of the tile and the placement point on the lattice (see Figure 6-9b). That is, given a lattice, one places the tiles in the grid (first approach), or place the tiles on the lattice points of that grid (see Figure 6-9). Mathematically, they are the same (if the lattice is regular). However, from a practical perspective they have different pros and cons. In the first approach, the interface would change as the wallpaper type changes but the same tile cannot be used in different wallpaper types, something that one might want to evolve in the tool. In the second approach, the interface and tile shape is always the same but it is harder to judge and control the overlap between tiles which can radically affect the design. When reflection and rotation are involved, it is hard to determine how the final wallpaper will look from a given tile.

However, there are existing tools (such as Artlandia's SymmetryWorks® (artlandia.com)) in which one is able to draw a shape on screen and immediately see that shape tiled into a wallpaper. Then one can move and otherwise modify any of the shapes in the wallpaper and those changes will immediately be reflected in the whole wallpaper. In the case of fabrics and wallpapers—importantly, where the end goal is a *tessellation* rather than a tile—a very powerful tool could be created by marrying such a wallpaper design tool with an evolutionary tool such as evoDesign whereby one would evolve whole fabric or wallpaper design at once rather than individual tiles.

References

Bandte, O., and Malinchik, S. A (2004) broad and narrow approach to interactive evolutionary design—an aircraft design example. In GECCO 2004 (K. Deb et al., eds), LNCS 3103, pp. 883–895. Springer Verlag, Berlin.

Bentley, P. J., and O'Reilly, U.M. (2001) Ten steps to make a perfect creative evolutionary design system. In the GECCO 2001 Workshop on Non-Routine Design with Evolutionary Systems.

Bentley, P. J. Aspects of Evolutionary Design by Computers. In Advances in Soft Computing - Engineering Design and Manufacturing, Springer-Verlag, London, 1999, pp. 99–118.

Dorris, N., Carnahan, B., Orsini, L., and Kuntz, L.-A. (2000) Interactive evolutionary design of anthropomorphic symbols. In Proceedings of Congress on Evolutionary Computing (CEC'04) (Portland, OR, June 19–23, 2003) IEEE, 433–440.

Eckert, C., Kelly, I., and Stacey, M (1999) Interactive generative systems for conceptual design: an empirical perspective. In Artificial Intelligence for Engineering Design, Analysis and Manufacturing 13, 303–320.

Feldman, U. (2005) Quantifying the dimensions of color experience. Unpublished Ph.D. thesis. Massachusetts Institute of Technology.

Frowd, C.D., Hancock, P.J.B., and Carson, D. (2004) EvoFIT: A Holistic, Evolutionary Facial Imaging Technique for Creating Composites. ACM TAP, Vol. 1 (1), pp. 1–21.

Kelly, I. (1999) An evolutionary interaction approach to computer aided colour design. Ph.D. Thesis, The Open University, Milton Keynes, UK.

McCormack, J. (1993) Interactive evolution of L-systems grammars for computer graphics modeling. In Complex Systems: from Biology to Computation (D. Green, T. Bossomaier, eds) ISO Press, Amsterdam.

Rooke, S. (2005) The evolutionary art of Steven Rooke. http://www.dakotacom. net/~srooke/index.html (Jan 7, 2005).

Sims, K. (1991) Artificial evolution for computer graphics. ACM Trans. Comput.Graphics 25, 319–328.

Takagi, H. (2000) Active user intervention in an EC search. In 5th Joint conference on Information Sciences (JCIS2000), Atlantic City, NJ.

Takagi, H. (2001) Interactive evolutionary computation. In Proceedings of the IEEE 89(9), 1275–1296.

Takagi, H. (1998) Interactive evolutionary computation—cooperation of computational intelligence and KANSEI. In Proceedings of the 5th International Conference on Soft Computing and Information/Intelligent Systems (IIZUKA '98), World Scientific, 41.

Todd, S., and Latham, W. (1992) Evolutionary Art and Computers. Academic Press.

Whitelaw, R. (1999) Breeding aesthetic objects: art and artificial evolution. In Creative Evolutionary Systems (edited by Bentley, P.J. and Corne, D.W). Morgan Kaufman. 129–145.

Whitelaw, R. (2004) Metacreation: Art and Artificial Life. Cambridge, Mass.: MIT Press.

Chapter 7

OPTIMIZATION OF STORE PERFORMANCE USING PERSONALIZED PRICING

Cem Baydar

Peppers & Rogers Group, Buyukdere Caddesi Ozsezen Is Merkezi No:122 C-Blok Kat 8, Esentepe, Istanbul, Turkey

Abstract   In this chapter, we discuss how to optimize a grocery store's performance using personalized pricing and evolutionary computation. Currently most of the grocery stores provide special discounts to their customers under different loyalty card programs. However, since each individual's shopping behavior is not taken into consideration, these discounts do not help optimize the store performance. We believe that a more determined approach such as individual pricing could enable retailers to optimize their store performance by giving special discounts to each customer. The objective here is to determine the feasibility of individual pricing to optimize the store performance and compare it against the traditional product-centered approach. Each customer is modeled as an agent and his/her shopping behavior is obtained from transaction data. Then, the overall shopping behavior is simulated and the store performance is optimized using Monte-Carlo simulations and evolutionary computation. The results showed that individual pricing outperforms the traditional product-centered approach significantly. We believe that the successful implementation of the proposed research will impact the grocery retail significantly by increasing customer satisfaction and profits.

Keywords: Customer modeling and simulation, agent-based simulation, one-to-one modeling, individual pricing, evolutionary computation

1.   INTRODUCTION

As the competition in retail industry increases, retailers are becoming much more obligated to optimize their store performance. For sectors with tighter profit margins and where customer loyalty is highly dependent on prices offered, it becomes more crucial to understand the customer behavior. Grocery retail is one of these sectors. Currently most of the grocery chains in the U.S

C. Baydar: *Optimization of Store Performance Using Personalized Pricing*, Studies in Computational Intelligence (SCI) **88**, 143–161 (2008)
www.springerlink.com

offer loyalty programs. It has been reported that 70% of all U.S households participate in some type of loyalty card program for grocery shopping. However, these loyalty programs mostly apply *blanket couponing* technique by offering the same discounts to their subscribers. This means that the information about each individual's shopping behavior is underutilized by aggregating and averaging the data and assuming that most people have similar product and price preferences. However, humans are different and each individual has his/her own preference of products and price levels. Therefore modeling each customer separately and providing him/her individual coupons could improve the store performance. This type of offering is known as *one-to-one marketing* in the literature.

The definition of one-to-one marketing is stated as the share of customer, not just market share (Allenby and Rossi, 1999). Therefore, instead of trying to sell as many products as possible over the next sales period, the aim is building loyalty with each customer by treating him/her as an individual rather than a part of a segment. Several approaches have been proposed for individual modeling in the literature (Dewan et al. 1999, Haruvy and Erev 2001). However, none of these approaches extended their work to optimize the store performance. We believe that it is possible to optimize the store performance by building individual models and testing pricing strategies using these models.

In order to realize the feasibility of individual pricing concept, let us assume that it is possible to record the shopping behavior of each customer of a grocery store - how often they come into the store, what they buy and how often, and which products they are especially price sensitive to. This information can be utilized by building a predictive model for each individual. Although this type of information is currently being collected at the checkout by grocery stores, it is not being utilized.

Our proposed approach assumes that by using a sufficiently rich transaction data, it is possible to capture each regular customer's shopping behavior. Then, individual models (agents) can be generated using this behavioral information and an agent-based system can be developed to simulate overall shopping behavior. The inputs for this agent-based simulation system can be provided by a store manager based on a strategy defined by the relative importance of three factors: *profits, sales volume* and *customer loyalty*. Finally, the system can use agent-based simulations in combination with evolutionary computation to identify the set of discounts for each customer. Figure 7-1 shows the overall approach.

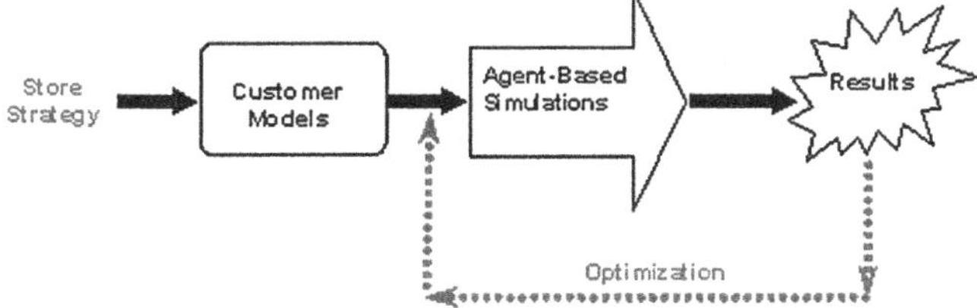

Figure 7-1. Outline of the proposed approach.

We have developed a system and tested the proposed approach against different blanket couponing pricing strategies. The results showed that individual pricing outperforms blanket couponing approach significantly. We believe that retailers can optimize their store performance by applying individual pricing.

2. PREVIOUS WORKS

One-to-one marketing is a customer relationship management paradigm which aims to build customer loyalty by trying to sell as many as products as possible to one customer at a time (Peppers and Rogers, 1999a and 1999b) Unlike the traditional *clustering approach*, one-to-one marketing aims to treat each customer as an individual rather than a part of a segment. Frequent flyer programs offered by airliners are one type of examples of this approach. There are also similar types of loyalty programs offered by on-line music retailers.

Grocery retail has always been an interest for the application of one-to-one marketing. The main advantage is that in grocery business almost every customer is a repeated buyer and grocery goods are consumed at a constant rate.

Therefore, there is sufficient amount of data to model each regular customer's shopping behavior.

In retail industry, most supermarkets use customer loyalty cards. Several companies have also started to analyze the premise of one-to-one marketing. For example Tesco has identified over 5,000 "needs segments" among its customers and improved its inventory management, product selection, pricing and discounts significantly using one-to-one marketing approach. Catalina Marketing has more than 1.8 terabytes of market basket data and is able to analyze market baskets thoroughly to obtain customer purchasing behavior. Eldat, which is responsible for manufacturing electronic shelf labels for retailers, has a product which can be used to convey personal pricing information to customers in a grocery store.

In one of our early works (Baydar, 2002a), we have discussed a conceptual framework for individual pricing and suggested that using agent-based simulations would be useful to analyze and optimize grocery store performance. Furthermore, we have discussed the preliminary results of an optimization algorithm which can be used for personalized pricing (Baydar 2002b, 2003). In this chapter, we will describe our implemented work and results in detail. Our results showed that one-to-one marketing outperforms the traditional blanket couponing approach significantly. We expect most grocery stores will start targeted couponing in the near future. This approach will even be more important if, in the future, the use of hand-held devices equipped with Bluetooth or other wireless connections will become widespread. Shoppers will use them to get personalized information about the products they see. In this context, dynamic individual promotion of products becomes a competitive necessity.

3.   PROPOSED APPROACH

Our approach uses an *agent-based* (Ferber, 1999) modeling and simulation approach which is different from the more focused store optimization research approaches found in the literature. In agent-based computational modeling, only equations governing the micro social structure are included (i.e., shopping behavior of each individual). Then, the overall macroscopic structure of the system grows from the bottom-up. Typically for grocery store optimization, revenues, costs and sales volume are taken into account as complex mathematical equations. However in agent-based approach, these values are determined by summing up each customer's shopping activity such as his/her shopping frequency and spending. The implementation steps of our approach are as follows:

1. Model each customer's shopping behavior from transaction data.
2. Create customer models as agents using these models.
3. Perform agent-based simulations and optimize the store performance for a given store strategy.

3.1 Problem Statement and Formulation

A grocery store manager has to decide on the store strategy based on the *relative* importance of three goals: *profits, sales volume* and *customer satisfaction*. These goals are contradictory (i.e., a store manager could maximize customer satisfaction by reducing all prices to zero). Therefore, what determines the overall store performance is the difference between each objective. We can visualize the task of setting a store strategy as adjusting the three levers as shown in Figure 7-2.

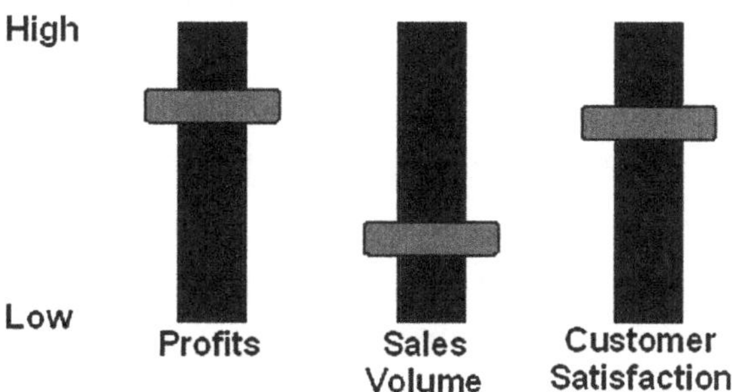

Figure 7-2. Three goals to determine store strategy.

The optimization strategy can be defined in mathematical terms as:

$$Maximize\ f(x, y, z) = w_1{}^*x + w_2{}^*y + w_3{}^*z \tag{1}$$

where;

x = profits
y = sales volume
z = customer satisfaction
and w_1, w_2 and w_3 are the appropriate weights determined by the store manager.

Since we are using agent-based models, there is no way of exploring x, y and z dimensions directly. Therefore, they are not the decision variables. The decision variables of this problem are the set of discounted products and discount values for these products. Both of these variables are different for each customer since we are giving individual discounts. Therefore, two questions are being addressed to maximize the objective function:

1. What is the optimal set of products for each customer?
2. What should be the discount values on these products?

It is also assumed that the shopping behavior of one customer does not affect another (i.e., shopping behaviors are not dependent). Therefore if we have n customers, this assumption enables us to write the objective function for each customer as:

$$i = 1, \ldots, n \tag{2}$$

$$Maximize\ f_i(x, y, z) = w_1{}^*x_i + w_2{}^*y_i + w_3{}^*z_i \tag{3}$$

Subject to:

$$P_i = \{P_1,\ P_2, \ldots P_r\} \tag{4}$$

$$D_i = \{D_1,\ D_2, \ldots, D_r\} \tag{5}$$

where;

i = Customer ID

n = Total number of customers

r = Size of the product and discount sets (i.e., number of coupons)

P = Set of products containing product ID's

D = Set of discounts containing the coupon face values

For a typical grocery store, there are about 1,000 customers and 50,000 to 80,000 products at store keeping unit (SKU) level. Our previous analysis showed that (Petrushin, 2,000) a typical customer buys nearly 300 different products a year. Even with this reduction, the size of our search space still stays large enough that finding the optimal discount values for each customer and requires a unique approach.

3.2   Problem Modeling

There are two types of models that we consider for this problem: *store model* and *customer model*.

3.2.1   Store Model.   The store model consists of several parameters such as:

- The number of products
- Quantity stored for each product
- Sales price of each product
- Product replenishment frequency
- Replenishment threshold
- Replenishment size
- Daily stock keeping cost of each product (inventory cost)

The parameters about replenishment determine the supply rate of a product. We assume that there is a constant supply frequency and replenishment amount when the quantity of a product is below a specific threshold. In addition to measure inventory costs, a daily stock keeping cost for each product is used.

3.2.2   Customer Model.   Each customer is modeled with several shopping properties such as:

- Shopping frequency
- Price sensitivity for each product
- Buying probability for each product
- Consumption rate for each product

Shopping frequency is modeled with parameters of first day of shopping (phase), frequency of shopping and probability of arrival at the expected day. For example, a customer may prefer shopping once a week on Saturdays with 90% probability. During the simulations, a uniform probability distribution function is used to sample the parameter of probability of arrival.

Price sensitivity is defined for each product since a customer may have different shopping behavior towards each product. For example, a person may prefer buying milk all the time regardless of its price but on the other hand he/she may be very price sensitive to beef. A person's buying probability can be influenced by giving a discount. This change is formulated as:

$$\Delta BP = (1 + \Omega(k * d)) \tag{6}$$

where;

ΔBP is the change in buying probability,
d is the discount rate,
k is the price sensitivity and
Ω is a probabilistic normal distribution function with mean and standard values as k^*d and $(1/3)^*(k^*d)$ respectively.

As it can be seen from Equation (6) above, change in shopping behavior due to discounts is probabilistic. Therefore Monte-Carlo simulations are used to obtain accurate results. The value k is obtained from the data for each customer using the following metric:

$$k = 1 - \left(\frac{\sum\limits_{i} p_i v_i - p_{\min} \sum\limits_{i} v_i}{(p_{\max} - p_{\min}) \sum\limits_{i} v_i} \right) \tag{7}$$

where;

k is the price sensitivity;
v_i is the amount of product purchased at i-th visit
p_i is the price of the product at time of i-th visit
p_{max} is the maximum price;
p_{min} is the minimum price.

The above formula suggests that if the customer has paid the minimal price for all purchases of the product, then his/her price sensitivity is 1; if he/she has paid maximal price, then his/her price sensitivity is 0. The following formula is used to calculate the updated buying probability:

$$BP(A) = BP'(A) * \Delta BP(A) \tag{8}$$

where;

BP(A) is the new buying probability of product A after price change.
BP'(A) is the buying probability before price change.
$\Delta BP(A)$ is the change in buying probability due to the discount offer.

In addition to these properties, there are two behavioral rules:

1. As the customer buys a product continuously, he/she starts building loyalty towards that product (i.e., buying probability increases).

2. If the customer finds the prices high for him/her or can not find a product
   from his/her shopping list, he/she gets frustrated and his/her probability
   of arrival decreases.

Figure 7-3 shows the product hierarchy used in modeling of the products as
suggested in an earlier work (Petrushin, 2000).

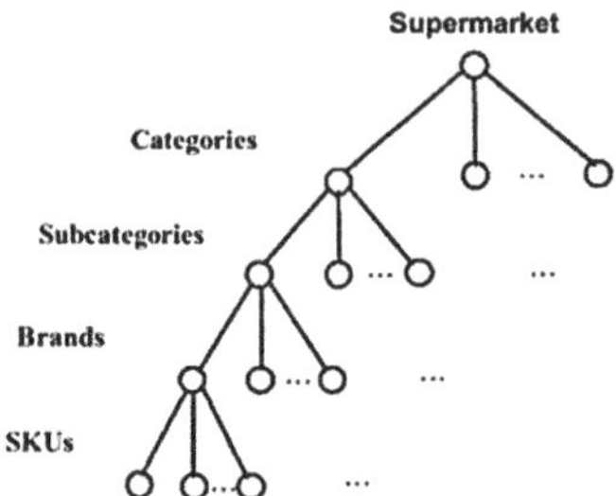

Figure 7-3.   Grocery store Inventory Representation

The inventory of a typical grocery store can be represented as a hierarchy of
categories (i.e., Dairy), subcategories (milk, egg, etc.), brands and store keeping
units (SKUs). For example "Country Fresh 2% Reduced Fat Milk" belongs to
the category "Dairy", subcategory "Milk", brand "Country Fresh" and has the
SKU #7160000901. A typical grocery store has 50,000–80,000 SKUs.

Understanding the associations between products is very important when
giving individual discounts. Each customer has an evoked set – the group of
products that a consumer would consider buying from among all the products
in the store hierarchy which he or she is aware and substitution or complemen-
tation can take place among this set. For one customer, Pepsi and Coke may
be substitutes but for another who likes both products, the two products may
be independent. If a discount is given on one of the substitute or complement
products, the other product's buying probability will also change. Two types of

association are possible between products: *complements* and *substitutes*. Complement products are found together in the basket (i.e., pasta and pasta sauce). Therefore, giving a discount on one product may also increase the buying probability of other product. On the other hand, substitutes tend to repel each other (i.e., Coke and Pepsi). For this type of products, giving discounts on both products may not affect the buying probability. Therefore in order to give individual discounts efficiently, product dependencies should be examined carefully.

One way of understanding whether two products are dependent is using a statistical dependency test. If two products are independent, the probability of their co-occurrence is the same as the product of the probabilities of the individual events. For example, if Coke and Pepsi occurred separately in 25% of all baskets, the expected co-occurrence of these two products is 6.25%. Any deviance (positive and negative) from this expected value may indicate product dependency. In order to determine the statistical significance of this deviance, the Chi-Squared statistical test can be used. The following formula is used for this purpose:

$$X^2 = \frac{[Expected\_ - Co - Occurance - Actual\_Co - Occurance]^2}{Expected\_Co - Occurance} \quad (9)$$

Note that X^2 measures deviation from expected random occurrence. The value can be compared with the X^2 table and confidence on the dependency can be obtained. Another suggested method is *product triangulation* (Castro, 1999), which is taking another product and comparing the dependencies of two products against this product to understand their association.

It is imperative that when giving individual discounts, the targeted products should be chosen carefully in order to obtain better store performance. Ineffective discounts may decrease both the customer satisfaction level and profitability. If there are two substitute products A and B, the buying probability of the dependent product B changes according to the given discount on product A using the following formula:

$$\Delta BP(B) = -\frac{BP(B)}{BP(A) + BP(B)} \cdot \Delta BP(A) \qquad (10)$$

As it can be seen from the equation above, if the change in the buying probability of product A is positive, the change in the substitute product is negative. The change is proportional to the relative importance of the buying probabilities between product A and B. For complement products, the change is directly proportional with product A, so the negative sign should be removed.

Finally, each customer has a satisfaction function. In order to measure this, we calculate the sum of the buying probabilities of the products which are expected to be purchased by the customer when he/she comes into the store. Then, we calculate the sum of buying probabilities of the products, which were bought in the simulation after discounts. The satisfaction function is defined as the ratio of these two summations as given in the following equation:

$$S.F = \frac{\sum BP_a}{\sum BP_e} \qquad (11)$$

where;

BP_a is the simulated buying probabilities after discounts,
BP_e is the expected buying probabilities

As discussed earlier, if a person can not find an item from his/her shopping list or finds the price high, he/she skips buying that product. Therefore, his/her satisfaction function decreases proportionally depending on the buying probability of that item (i.e., favorite items have much impact on the satisfaction function). This also affects his/her shopping arrival probability.

3.3   Optimization

The overall optimization stage is composed of 3 steps:

1. Performing sensitivity analysis on the product space of each customer to select the most suitable products from substitute pairs.
2. Applying the optimization algorithm shown in Figure 7-4.
3. Ranking of the products to identify the product set for a specified number of discount coupons.

Since discounts should be given on only one product from each substitute group, the first step is reducing the search space by selecting these suitable products. In this step, we pick products one-by-one from each substitute pair and perform sensitivity analysis by applying a 1% discount to that product. Then, we simulate the shopping behavior and compare the store performance in profits, sales volume and customer satisfaction between all substitute products. Based on these comparisons, the product which has the most effect on store performance is chosen from each product group. By following this procedure for each customer, we reduce the number of product space for the optimization phase.

In the second step, we apply the optimization algorithm to the set of products selected and obtain the optimal discounts to maximize the store performance. In order to solve this optimization problem, we have developed a hybrid parallel simulated annealing algorithm which uses the survival of the fittest method based on evolutionary computation concepts (Mahfoud and Goldberg, 1992). At first, the search space is divided into n equal parts and a population of m starting points is selected from each part. Then, using simulated annealing each member starts exploring its neighborhood in a parallel fashion. After each evaluation, better members are replicated while worse members are eliminated from the population based on their fitness value, which is the value of objective function, or in other words, the store strategy. It should be also noted that we evaluate the objective function $f(S)$, k times using Monte-Carlo simulation since the shopping behavior is probabilistic. This evaluation makes the problem computationally extensive. By eliminating worse members in the population, we also reduce unnecessary computations in the non-promising regions and explore the more promising regions with multiple members in parallel. Detailed information about this algorithm can be found in our previous work (Baydar,

2002b). Since the algorithm uses simulated annealing, there are three variables selected arbitrarily: These are the starting temperature (T_s), temperature reduction ratio (r) and cooling temperature (T_c). Figure 7-4 shows the flow chart of the algorithm.

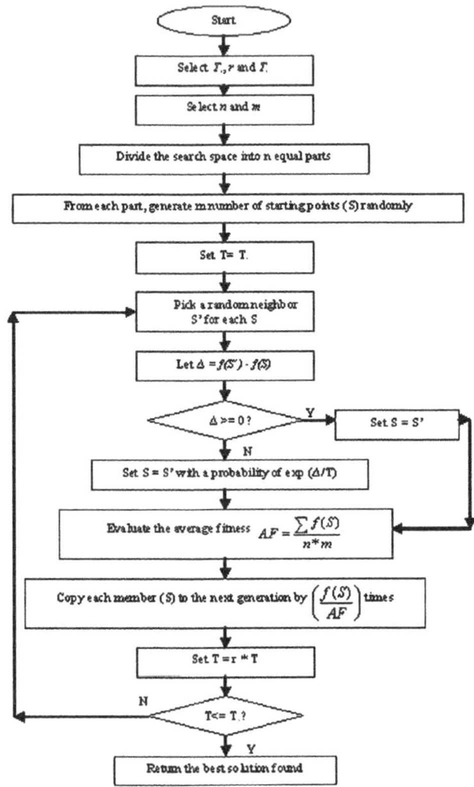

Figure 7-4.   Flowchart of the Optimization Algorithm

If the objective is to distribute a limited number of coupons, a ranking approach is applied. Similar to the first step, another sensitivity analysis is conducted and by selecting one product at a time, the discount value obtained in the second step is applied (i.e., the coupon is given). Then, the effect of this discount on store performance is evaluated by simulations. After that, each product's effect on store performance is ranked. Finally, the required number of products is selected from this sorted list. For example if 10 coupons are required, the top 10 products are selected for each customer.

4.　　CASE STUDIES

In order to compare the two approaches of personal pricing and traditional blanket couponing, we have built a sample database of 200 customers with 100 products and evaluated the performance difference against the same allowance on promotion spending. As a promotion strategy, for the following 15 days, we would like to spend \$1,150 on the discounts and we want to maximize the customer satisfaction.

The first approach is a traditional method, such as giving 10% discount on the top-10 favorite products. The second approach is by following the individual discounting strategy, giving 10 coupons to each individual at the store entrance with different discount levels on different products. For the optimization process we have selected our objective function as:

$$Maximize\ f(x,\ z) = 0.25^*x + 0.75^*z \qquad (12)$$

where;

x = profits and
z = customer satisfaction

Both approaches were simulated and the results showed that individual pricing outperforms the traditional approach significantly with a higher customer satisfaction: 8.75% vs. 3.35%. Figure 7-5 shows the results.

Customer Satisfaction Change

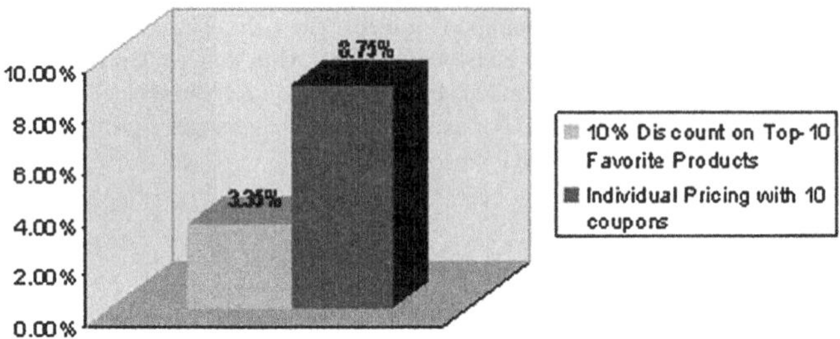

Figure 7-5. Results of the first case study.

In the second case study, we compared the amount of money needed on promotions for the same change in customer satisfaction. Now the question is: "As a store manager, if I want to increase the overall customer satisfaction by 3.35% how much should I spend on discounts?" This time we have selected the weights of our objective function as (0.75, 0, 0.25). Therefore, our objective is:

$$Maximize\ f(x,\ z) = 0.75^{*}x + 0.25^{*}z \qquad (13)$$

where;

$x =$ profits and
$z =$ customer satisfaction

For a traditional approach such as giving 10% discount on top 10 popular products, it costs \$1,150. However, with the individual pricing approach, the optimization results showed that the same amount of change can be achieved by giving five coupons this time and spending less than 1/3 of the traditional approach. The following figure shows the results:

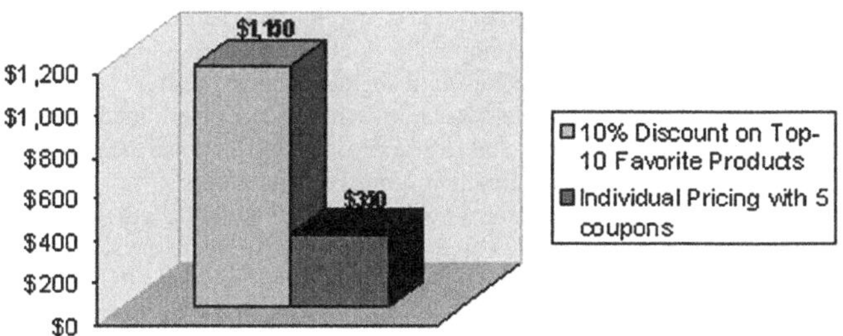

Figure 7-6.   Results of the second case study.

These case studies show that personalized pricing outperforms the traditional product-centric approach significantly with increasing customer satisfaction and profits. Although the numerical results on customer satisfaction and profits depend on the nature of the shopping behavior of the customer population, we believe that for other cases of personalized pricing will also outperform the traditional approach since it optimizes the store performance by looking at each customer's shopping behavior.

5.   DISCUSSIONS AND CONCLUSION

The continuously increasing competition in retail industry pushes companies into a position of communicating with their customers much more efficiently than they used to. For retail sectors with tighter profit margins where customer loyalty is highly dependent on the prices offered, it is essential to optimize the resources spent on increasing the customer satisfaction. Grocery retail is one of these sectors. Currently, most of the grocery stores provide a type of loyalty program which provides the same discounts to the subscribed customers. However this product-centered approach is efficient up to some level since customers are being divided into several segments and treated as a part of the segment rather than an individual. We believe that a more determined approach, such as individual pricing will enable grocery stores to increase their customer satisfaction levels without sacrificing too much of the profits.

Our discussed approach is based on agent-based modeling and simulation, which models each customer's shopping behavior to simulate the store performance. First, each customer is modeled using the available data sources. Then based on a given store strategy, simulations are performed to optimize the store performance by giving individual discounts.

We have developed a system to simulate the shopping behavior and optimize the store performance. We have conducted several case studies using this environment and compared the performance of two approaches. The results showed that individual pricing outperforms the traditional product-centered approach significantly. Several prototype-level implementations have been conducted with industry partners and encouraging results were achieved. We believe that the discussed approach will impact the grocery retail industry significantly by increasing customer satisfaction, sales volume and profits.

Acknowledgements

This work was completed when the author was with Accenture Technology Labs.

References

A.C. Nielsen's 4th Annual Frequent Shopper Survey. (2001), http://www.acnielsen.com

Allenby G.M., Rossi P.E. (1999), "Marketing Models of Consumer Hetero geneity", Journal of Econometrics, v89, pp.57–78.

Baydar C. (2002a), "One-to-One Modeling and Simulation: A New Approach in Customer Relationship Management in Grocery Retail", SPIE Conference on Data Mining and Knowledge Discovery: Theory, Tools, and Technology IV, Orlando 2002.

Baydar C. (2002b), "A Hybrid Parallel Simulated Annealing Algorithm to Optimize Store Performance", 2002 Genetic and Evolutionary Computation Conference, Workshop on Evolutionary Computing for Optimization in Industry, New York.

Baydar C. (2003), "Agent-based Modeling and Simulation of Store Performance for Personalized Pricing", 2003 Winter Simulation Conference.

Castro M. (1999), "Mining Transactional Data", Data Mining: A Hands-On Approach for Business Professionals, 2nd Edition, Prentice Hall.

Catalina Marketing Website, http://www.catalinamarketing.com

Dewan R., Jing B., Seidmann A. (1999), "One-to-One Marketing on the Internet", Proceeding of the 20th International conference on Information Systems, pp. 93–102.

Eldat Systems Website, http://www.eldat.com

Ferber J. (1999), "Multi-Agent Systems: An Introduction to Distributed Artificial Intelligence", Addison Wesley.

Haruvy E., Erev I. (2001), "Variable Pricing: A Customer Learning Perspective", Working Paper.

Mahfoud S., Goldberg D. (1992), "A Genetic Algorithm for Parallel Simulated Annealing", Parallel Problem Solving from Nature, 2, pp. 301–310.

Peppers D., Rogers M. (1997), "The One to One Future: Building Relationships One Customer at a Time", Double Day Publications.

Peppers D., Rogers M. (1999a), "Enterprise One to One: Tools for Competing in the Interactive Age", Double Day Publications.

Peppers D., Rogers M. (1999b), "The One to One Manager: Real-World Lessons in Customer Relationship Management".

Petrushin V. (2000), "eShopper Modeling and Simulation", Proceedings of SPIE 2000 Conference on Data Mining, pp. 75–83.

Rossi P., McCulloch E., Allenby G.M. (1996), "The Value of Purchase History Data in Target Marketing", Marketing Science, vol. 15, no. 4, pp. 321–340.

Tesco Website, http://www.tesco.com

Chapter 8

A COMPUTATIONAL INTELLIGENCE APPROACH TO RAILWAY TRACK INTERVENTION PLANNING

Derek Bartram[1,3], Michael Burrow[2] and Xin Yao[3]

[1]*Rail Research UK, The University of Birmingham, UK;* [2]*Railways Group, School of Engineering, The University of Birmingham, UK.* [3]*The Centre of Excellence for Research in Computational Intelligence and Applications (CERCIA), The University of Birmingham, UK*

Abstract   Railway track intervention planning is the process of specifying the location and time of required maintenance and renewal activities. To facilitate the process, decision support tools have been developed and typically use an expert system built with rules specified by track maintenance engineers. However, due to the complex interrelated nature of component deterioration, it is problematic for an engineer to consider all combinations of possible deterioration mechanisms using a rule based approach. To address this issue, this chapter describes an approach to the intervention planning using a variety of computational intelligence techniques. The proposed system learns rules for maintenance planning from historical data and incorporates future data to update the rules as they become available thus the performance of the system improves over time. To determine the failure type, historical deterioration patterns of sections of track are first analyzed. A Rival Penalized Competitive Learning algorithm is then used to determine possible failure types. We have devised a generalized two stage evolutionary algorithm to produce curve functions for this purpose. The approach is illustrated using an example with real data which demonstrates that the proposed methodology is suitable and effective for the task in hand.

1.   INTRODUCTION

There is a growing demand in many countries for railway travel, for example in the UK the demand for passenger transport has grown by over 40% in the past ten years (AoTOC, 2006). As this demand is not being matched by the increased building of new railway track there is a constant pressure to reduce travel times and to improve the reliability and efficiency of existing lines. This necessitates the use of faster vehicles with heavier axle loads and accelerates the

D. Bartram et al.: *A Computational Intelligence Approach to Railway Track Intervention Planning*, Studies in Computational Intelligence (SCI) **88**, 163–198 (2008)
www.springerlink.com

rate of railway track deterioration. As the railway system deteriorates over time it is necessary to maintain and renew components of the system periodically. Such maintenance and renewal costs may be considerable for a large railway network and in the UK, for example, the annual cost of maintenance and renewal was approximately £4 billions (Network Rail, 2006) in 2006.

In order to minimize expenditure on maintenance and renewal and at the same time maintain acceptable levels of safety, reliability and passenger comfort, it is necessary to have effective and reliable methods of predicting and planning railway track maintenance. These methods must analyze large amounts of data to enable current and future maintenance and renewal requirements to be determined. To facilitate this process a number of computer based systems have been developed for use in the railway industry as described below. The accuracy of these systems for maintenance and renewal planning however relies to a great extent on the availability of accurate, up to date, data and on expert judgment. This chapter describes an alternative approach based on a number of computational intelligence techniques, including evolutionary computation techniques, which are less reliant on the quality of the data and engineering judgment.

In Section 2, data for prediction and planning are discussed. Section 3 summarizes the current railway track procedures for maintenance management. This includes the data generally available to the permanent engineers, a brief explanation of the methodology adopted by existing decision support systems and their inherent shortcoming. In Section 4, our proposed system, a new approach to decision support systems design is outlined together with with important issues facing the system and various ways that these issues may be overcome. Section 5 presents the prototype implementation, an actual implementation of the proposed system, illustrating its various functions. Finally, Section 6 gives our concluding remarks.

2.   DATA FOR PREDICTION AND PLANNING

Conventional railway track combines materials, such as the rail, rail pads, sleepers, ballast and sub-ballast, in a structural system (Figure 8-1). This system is designed to withstand the combined effects of traffic and climate to the extent that, for a predetermined period, the subgrade is adequately protected and that railway vehicle operating costs, safety and comfort of passenger are kept within acceptable limits (Burrow et al., 2004). As the system's components deteriorate over time, measures of the condition of each of the components are collected to help make decisions regarding their maintenance and replacement (renewal). Typically data are collected on the mechanical condition of the ballast, fasteners and sleepers. Additionally data are collected on the number and types of rail failures (cracks and breaks), rail wear and rail corrugation (short wavelength defects which occur on the surface of the rail), and on the geometry of the track. The latter is the most widely used measure of the condition of the track for planning track maintenance and renewal and is described in more detail below.

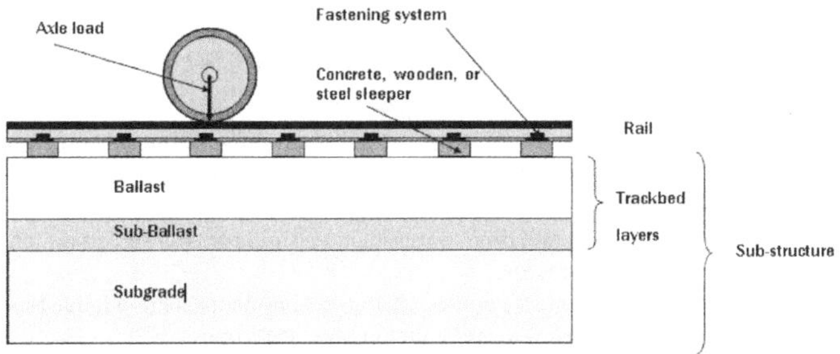

Figure 8-1.   Simplified Components of Conventional Ballasted Railway Track (Burrow et al., 2004)

2.1   Geometry Measurements

A large amount of railway track maintenance expenditure results from adjusting the position and amount of the ballast under the track to correct the line and level. When the ballast can no longer be adjusted to maintain adequate geometry it is replaced (renewed). However, the process of determining the type, amount, location and timing of such maintenance or renewal activities is a complex task which requires track geometry deterioration to be predicted (Esveld, 2001). As sections of track may have dissimilar rates and mechanisms of deterioration, track geometry data are periodically collected on short sections of track (typically less than 200m). The principal measures of track geometry collected are changes in the vertical and horizontal geometry and the track gauge over time. Such measures are usually obtained by vehicles, known as track recording cars, details of which may be found elsewhere (Esveld, 2001; Cope, 1993).

2.2   Work History Data

In addition to collecting data on the current condition of the track, historical data in terms of the type of maintenance and renewal work carried out at specific sites are usually recorded over a railway network. Such information may be used to help plan future maintenance and renewal activities.

2.3   Railway Standards

A number of railways have issued standards related to the measures of track condition described above. These are used to help formulate rules regarding maintenance and renewal requirements (see below).

3.   CURRENT RAILWAY TRACK PROCEDURES FOR MAINTENANCE MANAGEMENT

As described above, managing track maintenance effectively and efficiently is a complex task. To address this, a number of computer-based systems have been developed to deal with the large amount of data collected. The majority of these are rule based systems and aim to produce information on the most appropriate type, location and time of maintenance and renewal of components through optimization processes.

Rule based systems typically consist of four components, a knowledge base, a fact base, an inference engine and a diagnostic (see Figure 8-2).

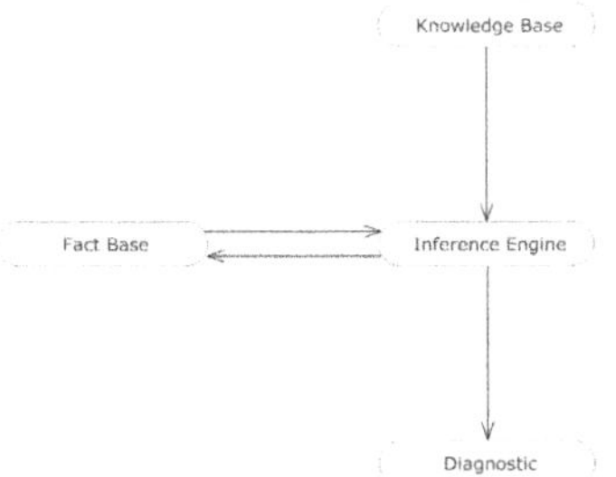

Figure 8-2.   Rule Based System Structure

In railway engineering, the knowledge base can be considered to contain information describing the type of components, physical location, age and associated measures of condition (Roberts, 2001b; Leeuwen, 1996). In addition the knowledge base may contain information regarding the maximum permissible speed of each section of track, the annual tonnage of the trains using each section, as well as a record of historical maintenance. In some systems, information from the knowledge base is used to determine deterioration models for each component. These models estimate the rate at which a component may be expected to deteriorate and are used to help plan future maintenance based on the expected condition of a component. The fact base contains rules which are processed by the inference engine and used in combination with information in the knowledge base to form a diagnostic in terms of the type, timing and location of maintenance or renewal activities.

Rules are typically of the following form (ABStirling99)

IF $(Condition_A) \ldots$ AND $\ldots (Condition_B) \ldots$ OR $\ldots (Condition_C)$
THEN *maintenance type and date*

For example (ERRI, 1996),

IF *(Rail is of Type A)* AND *(Speed $\leq 160km/h$)*
AND *(vertical wear of rail head in the track $\geq 8mm$)*
THEN *rail renewal required now*

While rule based systems are widely used, they have some inherent disadvantages, including;

- The quality of diagnostic is dependent upon the quality of the fact base and any errors or omissions in the fact base will result in the diagnostic being incorrect. The accuracy of the fact base are to a large extent based on the judgment of the engineers who have formulated them. Erroneous rules may result in a number of problems including maintenance and renewal work being scheduled early (and hence not being cost effective), scheduled late (and cause the track to fail to meet the comfort or safety specifications), or be of an incorrect type (which may result in only the symptoms of the underlying fault being treated). Furthermore, since the interrelation between the deterioration of all components in the track system is not fully understood, it is difficult for an engineer to define accurate rules which take into account these relationships. i.e. it is conceptually problematic to formulate rules which consider interactions between the rail, sleeper and ballast.

- As mentioned above, condition data are used to determine deterioration rates and models for components in isolation (ERRI, 1996). These models are often based on linear regression and represent a simplification of reality.

- Existing decision support systems are static solutions. Static solutions, once implemented, will always give the same result no matter how many data are processed by the system. Whilst static solutions may be dependable due to their predictability, as the quality of the system is dependent on the quality of the fact base, once an error is introduced into the system the error will always exist.

4. OUR PROPOSED SYSTEM

This section describes a prototype system which uses computational intelligence and data mining techniques to produce a dynamically generated output.

The system determines the diagnostics purely from historical data, and therefore overcomes the potential problems caused by domain expert knowledge (i.e. errors in the fact base). However, where there are insufficient historical data to enable an effective treatment to be suggested by the system, input from a permanent way engineer may be required. Using a system which derives its functionality from the data allows the system to be re-trained once more data have been acquired and hence, in theory, provide improved results each time new data are added. However, one problem using such a technique is that the initial resultant behaviour may be unpredictable. Furthermore, as the system considers all measures of track condition together, rather than in isolation as is the case for the existing systems, it is able to take into account any interactions between the various track components. Consequently, the maintenance and renewal plans determined by our proposed system are likely to be more accurate than that produced by systems which are unable to consider component interaction.

4.1   System Overview

The proposed system performs two tasks; training and application. In the training stage, a number of processes are carried out as described below;

1. Various possible types of failures (including failures which behave differently according to usage, track design and subgrade conditions) are determined. It should be noted that these failure types may not be the same as those traditionally identified by an engineer since the system considers the interactions between components.

2. For the failure types identified above, deterioration models are produced for each of the recorded geometry measurements.

3. Determination of the various intervention levels. This process may be omitted as the values calculated should be the same as those given in the railway standards, however for ease of use and setup this process has been included.

4. For each failure type the most suitable type of maintenance or renewal is determined.

5. Training of a classification system to categorize new track sections into the previously discovered failure types.

The system needs only be trained once before use, however for the outputs of the system to be improved over time, it should be periodically retrained as new data are added using the procedures described above. Retraining allows new failure types to be discovered and hence new deterioration models to be determined, thus improving the accuracy of the system outputs. Furthermore, as more data are added the deterioration models for all failure types can be regenerated increasing their accuracy.

Once the system has been fully trained, the following results can be produced;

1 The current failure type(s) of the track

2 The optimal time to the next intervention. This is achieved by using each of the deterioration models for the determined failure type in conjunction with the standards to determine the geometry measurements that will exceed the standards

3 The most suitable intervention type for the determined failure type.

4.2   Missing and Erroneous Data

In the railway industry during data capture it is possible to loose data for a number of reasons. These include; equipment error (incorrect setup or use) and measurement recorder out of bounds (particularly where the value measured is smaller / larger than the recorder can measure). Additionally, errors may be introduced in geometry recordings due to the limited accuracy of the recording tools used.

Figures 8-3 and 8-4 show two different sections of track and the corresponding values of a measure of track quality, known as the *standard deviation* of vertical top height, (y axis) against time (x axis). As can be seen in both examples, no work history data are available (typically this would be shown as a vertical line separating two successive measurements). In such a case the data must be pre-processed as described below in order to make it suitable for further analysis. Figure 8-4 shows the extent of noise within the data, which appears to become ever larger in magnitude. From observation of a number of deterioration patterns it would appear that the noisy sections are periodic in nature.

From the above it is evident that in order to produce a data driven approach to the problem described herein, three problems related to erroneous or missing data must be addressed as follows;

- Missing values

- Missing work history data

- Data noise.

These problems and their potential solutions are described below.

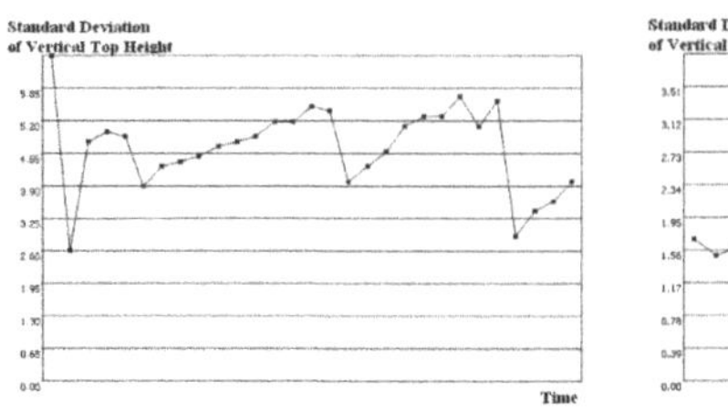 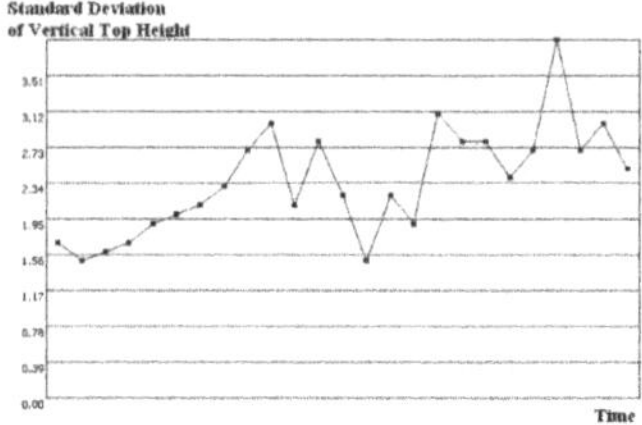

Figure 8-3.   Real Data Run (1)   *Figure 8-4.*   Real Data Run (2)

4.2.1   Handling Missing Values.   We investigated five different techniques to handle data with missing values:

- Removal of geometry measurements with a significantly high number of missing values removes any geometry measurement where the percentage (or otherwise) of missing values for the measurement exceeds some critical value. This technique is very good for removing measurements specified as being recorded, but have not been in practice. However, when the number of missing measurements is significantly high in general (i.e. a high proportion of missing values across all geometry measurements), the technique performs poorly as too much data are discarded.

- Removal of geometry records with a significantly high number of missing values discards records of geometry measurements with more than a critical number of missing values. This technique, like technique 1, has the disadvantage that a large amount of data are discarded from datasets with a significantly high number of missing values.

- Fill with a set value uses a predefined number inserted wherever a value is missing. This technique has the advantage of low complexity, however it will not perform well especially when an unsuitable value is chosen as the predefined number (e.g. a value outside the typical range for the measurement).

- Filling with a generated number is similar to filling with a set number but the number substituted is generated using an appropriate mathematical function, from the data itself. This technique has the advantage over substituting a predefined value as an appropriate function can be selected to generate missing values ensuring that the generated values are within the range of all possible values. In addition, it does not bias the data.

However, when it is necessary to replace values in a time sequence of data, the technique can cause an atypical peak or trough to occur in the processed data when the missing value is not near the median. An example is shown in Figure 8-5 where the mean value is used to replace the missing data. However, as the missing value occurs very early on in the deterioration cycle the replaced value is inappropriately high and if used would suggest that the track condition has improved prior to any maintenance treatment.

- Replace missing values with a run generated value, fits a curve (see Section 4.4 for curve fitting) to existing time series data to determine the missing value as shown in Figure 8-6 below. Even where the amount of data are limited, it is still possible to produce a more appropriate substitution using this technique than using one of the four methods described above. From the expected nature of track deterioration observed in practice and described in the literature (Roberts, 2001b), a third order curve function may be suitable for this task (since it can capture both the post-intervention and pre-intervention increased deterioration rates). However, since the amount of data in any one particular run are often limited, for the more generalized case, a linear function may be used instead. While the example in Figure 8-6 shows excellent missing value substitution, it is worth noting that the accuracy of the technique is dependent on the order of curve chosen for the curve fitting routine. For example, if the missing value is at the end, or beginning of a run and the curve shows a high degree of curvature due to an increasing deterioration rate pre and post intervention, using curve fitting of a higher order curve function is required. However, if too high an order curve function is used (i.e. over-fitting) the value chosen for substitution may be outside the bounds of values found in practice (see Figure 8-7).

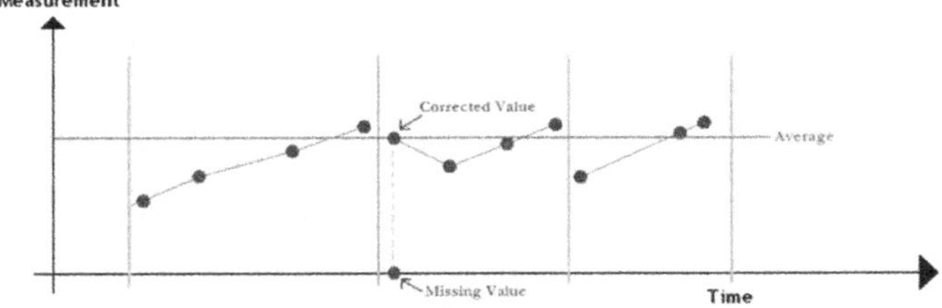

Figure 8-5.   Missing Value Filling Via Generated Value

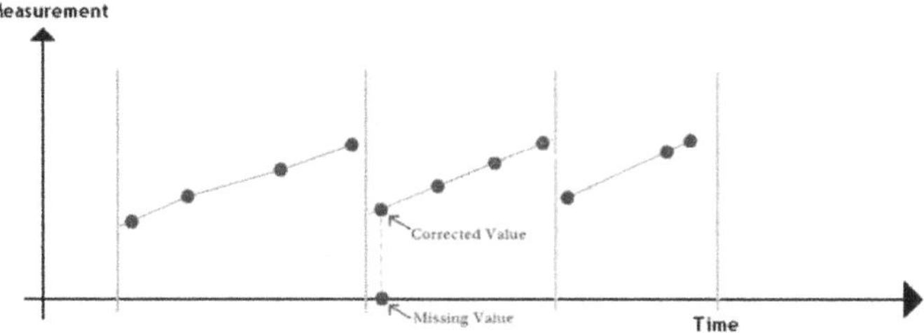

Figure 8-6. Missing Value Filling Via Run Curve Fitting

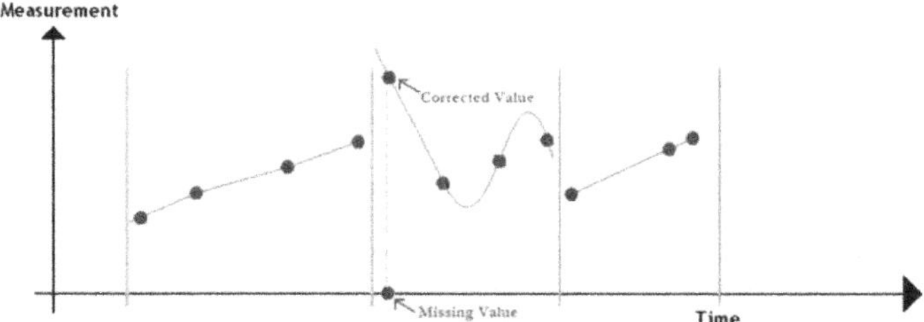

Figure 8-7. Missing Value Filled Via Poor Run Curve Fitting

4.2.2   Handling Missing Work History Data.   In some datasets work history data may be missing which therefore prevents the generation of runs of data from sets of data. As a result it is not possible to determine deterioration models from the data. However, as the maintenance work may be expected to increase the quality of the track, track quality data should indicate an improvement when maintenance or renewal work has been carried out. Consequently, in many such cases it is possible to determine from the track quality data alone when maintenance or renewal work has been performed. This is demonstrated in Figure 8-8, where it may be seen that distinct improvements in track geometry quality have occurred (Figure 8-9). It is also interesting to note that in the example shown in Figure 8-9 using such information, it is possible to surmise further and determine whether the intervention was maintenance or renewal work. Research suggests that after renewal the quality should be significantly

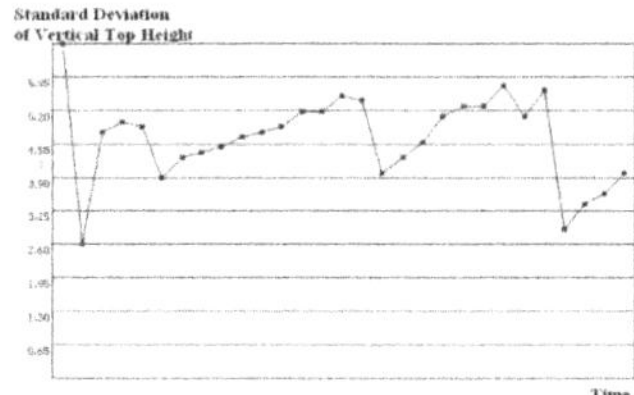

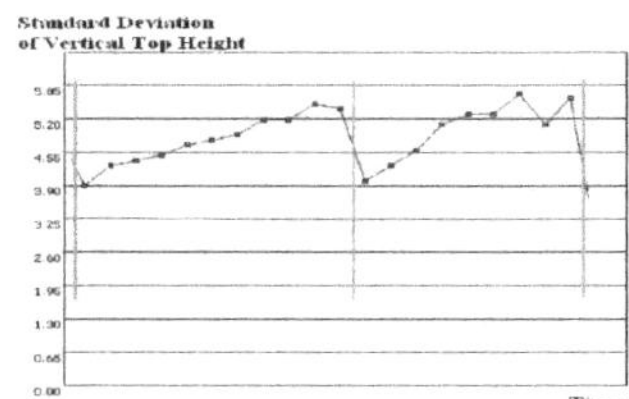

Figure 8-8. Standard Deviation Of Vertical Top Height By Time, Without Work History Data

Figure 8-9. Standard Deviation Of Vertical Top Height By Time, With Work History Data

better than after maintenance (Roberts, 2001b). Accordingly with reference to Figure 8-9, the first and last interventions are likely to be renewals, while the middle two are more likely to be maintenance activities. Furthermore, the characteristics of a typical deterioration curve reported in the literature can also be seen from the figure. Just before and also after intervention the deterioration rate increases. Additionally, due to the effects of maintenance on the long term performance of the railway track after each maintenance activity the average deterioration rate over time increases.

4.2.3 **Handling Noisy Data.** As shown in Figure 8-4, data may have a high degree of noise. To reduce its occurrence noise reduction techniques were used at two stages in the system training; once during the initial data processing stage (i.e. when the data runs are initially produced), and again later during the production of the deterioration models (during sampling for the genetic algorithms, see Section 4.4.1, Deterioration Modelling). The noise reduction techniques use a similar curve fitting technique to that described above (see Section 4.2.1), and fit a third order curve to each run. During the process it is necessary to ensure that sufficient data are available within the geometry time series so that the curve is an accurate representation of the data. Once a curve function has been determined for a run of data, the real data are scaled to match the curve in one of three ways; complete (so that all geometry points lie on the curve function upon scaling), partial (so that all geometry points lie some percentage of the way towards the curve function), or graduated (where points with a higher deviation from the curve function are scaled more or less than those with less deviation).

Figure 8-10 shows a set of data from two runs which have had the noise reduced (i.e. dark grey data to light grey data). In this example the data have been partially scaled to the curve function using a high scaling rate (i.e. a partial scaling with a relatively high percentage rate towards the curve function). Note however that the resultant data run does not necessarily show the expected trend of increasing in rate just prior and after intervention.

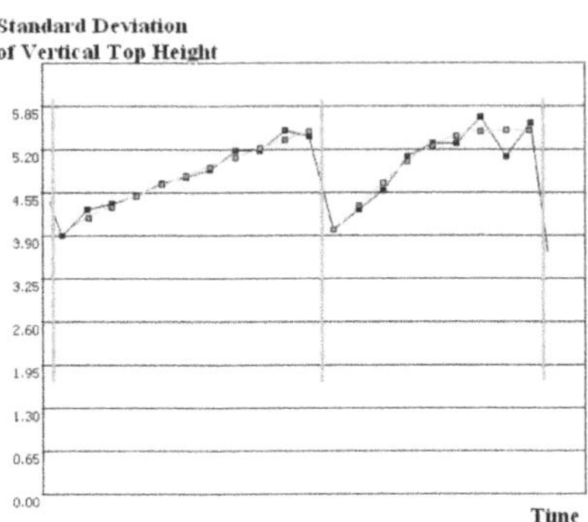

Figure 8-10.   Real Data Before And After Noise Reduction

4.3   Failure Types

One of the main tasks in developing the proposed system is the determination of failure types. For the purposes of the proposed system, faults which deteriorate differently for different track components are considered to be separate failure types.

For the task two common clustering techniques, K Means (Moore, 2004) and Rival-Penalized Clustering Learning (RPCL) (King and Lau, 1999), were considered and evaluated for their effectiveness. For both techniques a cluster centroid approach was adopted. In such an approach a point in n-dimension space represents the cluster and any geometry measurement (plotted within the same n-dimension space) is a member of the closest cluster centroid. Each cluster centroid therefore represents a single failure type.

For the work described herein each geometry measurement is assigned to a single dimension of n-dimension space and the clustering algorithm clusters within that space.

4.3.1   K Means Clustering.   K-Means is expressed in pseudo-code as follows (Moore, 2004);

- D is a dataset of i member elements $d_1, d_2, d_3...d_i$

- C is the set of n cluster centroids $c_1, c_2, c_3...c_n$, where n is predefined by the implementation

```
foreach (cluster_centroid c in C) {
    initialize c to a random position
}

do {
    foreach (dataset_element d in D) {
        assign d to the nearest cluster centroid
    }

    foreach (cluster_centroid c in C) {
        update c's position to mean of all elements
          previously assigned to it
        unassign all elements from c
    }
} until (all cluster_centroids in C have not changed
            position since previous iteration)
```

For the following sample dataset (Figure 8-11), the algorithm proceeds as follows;

1 From problem and domain knowledge, 2 clusters are known to exist, so 2 cluster centroids are initialized to random positions (Figure 8-12)

2 Data elements are assigned to the nearest cluster centroid (Figure 8-13)

3 Each cluster centroid is moved to the mean point of the average of all the points assigned to it (Figure 8-14)

4 Data elements are reassigned to the nearest cluster centre (Figure 8-15)

5 Cluster centroids are again moved to the centre of assigned data elements (Figure 8-16)

6 Data elements are reassigned to their nearest cluster centroid, however no elements are reassigned, and therefore clustering is complete.

In the proposed system as mentioned above, it is not desirable for the railway engineer to have to specify the failure types nor their number, since it is possible that they may not identify certain failure types. Therefore in the proposed approach it is assumed that the number of clusters is not known and consequently it is probable that K Means will not perform well. Since the number of failure types is unknown an estimate must be used, however using an unrealistic estimate will cause problems. If the estimate is too high, then some clusters will

Figure 8-11.   Sample Clustering Dataset   *Figure 8-12.*   K Means Clustered Dataset (1)

Figure 8-13.   K Means Clustered Dataset (2)   *Figure 8-14.*   K Means Clustered Dataset (3)

Figure 8-15.   K Means Clustered Dataset (4)   *Figure 8-16.*   K Means Clustered Dataset (5)

Figure 8-17. Over Estimate of Number of Clusters in K Means Clustering

Figure 8-18. Under Estimate of Number Of Clusters in K Means Clustering

be unnecessarily sub-divided (Figure 8-17), on the other hand an underestimate of the number of clusters can result in clusters being incorrectly joined(Figure 8-18).

4.3.2   Rival Penalised Competitive Learning.   Rival-Penalized Clustering Learning (RPCL) is a modified version of the K-Means algorithm which uses a learning and de-learning rate to remove the need for prior knowledge of the number of clusters. However the learning and de-learning rates require careful tuning to the size and density of the required clusters (King and Lau, 1999). In essence, RPCL performs similarly to K-Means with three modifications:

- Processing is performed on random data elements within the set, rather than all data elements nearest to the cluster centroid.

- Rather than moving cluster centroids to the mean position of assigned nodes, the closest cluster centroid is moved toward the randomly chosen point using a user specified learning rate.

- Unlike K-Means, RPCL moves the second closest cluster centroid (the rival) away from the randomly chosen point using a user specified de-learning rate.

RPCL is expressed in pseudo-code (King and Lau, 1999) as follows;

- D is a dataset of i member elements $d_1, d_2, d_3 \ldots d_i$

- C is the set of n cluster centroids $c_1, c_2, c_3...c_n$, where n is set to a value guaranteed to be greater than the actual number of clusters

- Learning rate 1_l is predefined in the range $0 < 1_l < 1$

- De-learning rate 1_d is predefined in the range $0 < 1_d < 1$, and typically $1_d < 1_l$

- Stopping criteria can be defined to best suit the domain, however typically they are a predefined number of iterations, or a reduction of learning and de-learning rates during each iteration until the learning rate or de-learning rate reaches a minimum bound.

```
initialize cluster centroids to random positions

while (stopping criteria not met) {
   pick random data element d_r from D

   determine closest cluster centroid c_c to d_r
   determine second closest cluster centroid c_r to d_r

   c_c.position = c_c.position + 1_l (d_r.position - c_c.position)
   c_r.position = c_r.position - 1_d (d_r.position - c_r.position)
}
```

For the following sample dataset (Figure 8-19) the algorithm proceeds as follows;

1. The number of clusters is unknown, but 3 was considered to be sufficient for this work (Figure 8-20). The literature varies on the best way of determining a sufficient number (Cheung, 2004), some favour selecting a very high number of clusters, while other suggest incrementing the number of clusters through repeated runs of the clustering algorithm until the number of output clusters does not change.

2. A random data element is chosen, and the winner (the grey cluster centroid) is learned towards it, and the rival (the white cluster centroid) is de-learned away from it (Figure 8-21).

3. The learning and de-learning rates are decreased. A new random element of data is chosen and the winner (the grey cluster centroid) is learned toward the point, and the rival (the black cluster centroid) is de-learned away from the point (Figure 8-22).

4. After a number of iterations the grey cluster centroid will tend towards the centre of the cluster, and the white and black cluster centroids will move

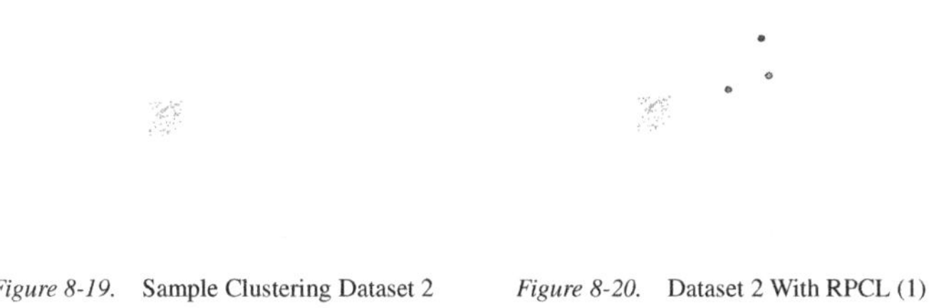

Figure 8-19. Sample Clustering Dataset 2   *Figure 8-20.* Dataset 2 With RPCL (1)

Figure 8-21. Dataset 2 With RPCL (2)   *Figure 8-22.* Dataset 3 With RPCL (3)

further and further away (indicating they are unneeded). Depending upon the finishing criteria, the algorithm may finish at this point or continue. Should the algorithm continue at this point grey continues towards the centre of the cluster and the white and black cluster centroids continue towards infinity.

RPCL however has a disadvantage, associated with the relative size of the clusters, which is of a particular concern when used in the railway intervention planning domain. In the examples given above, the clusters were all of the same size, however when the clusters are formed from geometry measurements representing failure types the clusters are unlikely to be of a similar size since some failures are more common than others. Using RPCL, points are selected at random, therefore the cluster centroids of the smaller clusters will be selected at random less often than the next nearest clusters which are likely to be significantly bigger. The smaller clusters will thus, on average, be learned away more than they will be learned towards the smaller cluster.

In the proposed system the problem of setting the learning and de-learning rate was addressed by first using the algorithm on a training set of data with known numbers of failure types and members. The training set was generated specifically for this task using ranges derived from actual data. The RPCL algorithm was used with different values until a good match between known properties of the training set and RPCL output was found.

4.4 Deterioration Modelling Using Evolutionary Approaches

For each geometry measurement of each failure type, a deterioration model must be produced. Unlike existing decision support systems which use a fixed mathematical function fitted to the available data for a particular data set, the proposed system uses a two step genetic algorithm to learn a curve function. A second genetic algorithm is subsequently used to fit the curve function to the data.

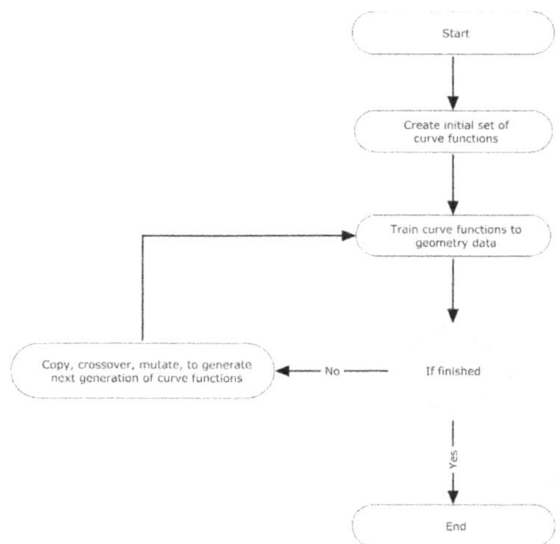

Figure 8-23. Deterioration Model Generation Flowchart

Using a two stage genetic algorithm provides several benefits. The main benefit is that the algorithm is able to generate a function automatically that will be able to model a typical deterioration curve, as well as any unusual deterioration patterns not normally considered. Furthermore, the genetic algorithm can use any standard mathematical function in conjunction with the known time since the last intervention, component ages, and traffic load. Since the curve function is built dynamically, it is also necessary to use a dynamic curve fitting function (i.e. a curve fitting function that fits to any curve, rather than a specific subset of curves). Figure 8-23 gives the deterioration model generation flowchart. Each step of the modeling process is explained using an example in the following subsections.

4.4.1 Deterioration Modelling - An Example.

1 Generalized curve functions are produced using random (but valid) combinations of the standard mathematical functions (+, -, *, /, Sin, Cos, Tan, ^ (power), etc), input variables (calculated from the data; time since last intervention and time since last replacement for each of rail, ballast, and sleeper), and algebraic constants (defined as 'a', 'b', 'c', etc). Constants are only defined in one place per formula; however two constants may obtain the same value. To improve the performance of the algorithm, the selection of the various formula components are skewed towards a third order polynomial since this is the likely function format. The following illustrates this process for a fictional dataset where G_m = ballast thickness.

$$G_m = \frac{ballast\_age}{a} + b^2 \tag{1}$$

$$G_m = ballast\_age^2 + b + \sin(c) \tag{2}$$

2 Train curve functions of population to data. Each formula within the current population is trained one at a time. To reduce complexity however, the sampling of the dataset may be pre-computed so that the sampling is only performed once per formula generation, rather than multiple times per generation.

    (a) If the dataset for training is sufficiently large (e.g. over 10,000 geometry records), a random sample (approximately 1,000) is taken to reduce training time while still producing a good match to the original data. For each geometry record all input variables are calculated (i.e. ballast age since last intervention, ballast life span, rail age, etc.). Table 8-1 shows example data used for deterioration modelling.

    (b) An initial random generation of constant values are generated. Assuming Formula 1 is being trained, then the following constant values may be generated (see Table 8-2).

    (c) The generation is evaluated using the following metric.

```
foreach (data row in training set) {
    calculate the difference between the formula
    output and desired result (i.e. geometry
    measurement)
}
```

Table 8-1.　Example Training Dataset For Deterioration Modelling

Ballast Thickness	Ballast Age (days)	Ballast Life Span (days)	Rail Age (days)	Etc...
25	15	15	1042	...
27	30	30	1057	...
28	42	42	1069	...
27	32	32	1059	...
29	48	48	1075	...
25	17	17	1044	...

Table 8-2.　Generation 0 Values For Constant Training

a	b
1	6
2	3.7
2	1
1.5	8
2	3.2

Table 8-3.　Metric Results, Generation 0

a	b	Metric
1	6	39.83333
2	3.7	4.396667
2	1	10.5
1.5	8	57.61111
2	3.2	4.086667

```
calculate average of all differences
```

Using this metric requires trying to minimize the function and results in the output shown in Table 8-3.

(d) Test end criterion; for this example, the end criterion is to complete 2 generations (i.e. generations 0 and 1), and therefore is not met.

(e) The next generation is formed using the current generation. The constants are converted to 32bit binary representations and joined. Subsequently, the next generation is formed via the standard techniques of crossover, mutation, and copying, with tournament selection. Once the next generation of 32bit binary representations are formed, the reverse process of concatenation is applied resulting in the outputs shown in Table 8-4.

(f) The generation is evaluated using the curve function evaluation metric as above, resulting in the output given in Table 8-5.

(g) Test end criterion; in this test, two generations have been completed, and so the following formula is returned back to the primary genetic algorithm.

Table 8-4. Generation 1 Curve Fitting Constant Values

a	b
2	3.15
1.9	3.15
1.9	3.35
2.15	3.4
2.15	3.3

Table 8-5. Metric Results, Generation 1

a	b	Metric
2	3.15	4.1925
1.9	3.15	4.174956
1.9	3.35	4.100482
2.15	3.4	3.669922
2.15	3.3	3.893256

$$G_m = \frac{ballast\_age}{2.15} + 3.4^2 \qquad (3)$$

3 Once all formulae in the generation are trained, the end criterion is tested. The end criterion for this example is the fitness as evaluated by the secondary genetic algorithm and is less than 3.5. In an actual implementation this may be changed to a more complex metric and may include components such as the minimum number of generations and the change in the best, or average generation fitness. Equation 1 returned a best fitness of 3.669922, while equation 2 returned a best fitness of 12.387221; therefore the end criteria is not met and so the primary algorithm continues.

4 The next generation of formulae is generated by the primary algorithm, using standard tournament selection and crossover, mutation, and copying. For example, Equation 1 is mutated to equation 4, and equation 1 and equation 2 are crossed over to form equation 4.

$$G_m = \frac{ballast\_age}{a} + b \qquad (4)$$

$$G_m = \frac{ballast\_age^2}{a} + b^2 \qquad (5)$$

5 Generation 1 is trained to data using the process described in 'Train curve functions of population to data' previously. Since generation 0 was shown in detail, generation 1 will not be shown.

6 Test the end criteria; in generation 1, equation 4 returns a best fitness of 2.341 and therefore the primary genetic algorithm returns.

4.4.2   Deterioration Modelling — Optimization.   The process of deterioration modelling using a two stage genetic algorithm is highly computationally complex, and as a result, it takes a long time to complete. Consequently, two solutions to reduce the training time required were used as described below.

1 A distributed system (Coulouris et al., 2000) was used to enable more computers to process the task. Each instance of the secondary algorithm was assigned to a task, and sent to a single PC on a grid of computers. Consequently, as many instances exist per generation of the primary genetic algorithm (one per population member), the computation time was significantly reduced.

2 As can be seen from Figure 8-24, after the initially fast improvement of quality, the rate of improvement decreases dramatically. Using this knowledge it is possible to set the secondary genetic algorithm to a short number of generations and obtain an approximation of the obtainable metric value. Running the secondary algorithm for a shorter period of time reduces the overall time for computation significantly. However, to maintain the quality of the output before the primary genetic algorithm terminates, the final output must be retrained to the original metric using the secondary genetic algorithm.

Figure 8-24.   General Quality Metric Per Generation Trend

4.5   Works Programming

Works programming is the process of determining the type of intervention to perform given the type of failure. Traditionally, works programming is performed by a railway track engineer, however given the complexity of the domain of railway track intervention planning (i.e. the various components, interactions, materials, etc.) it is very difficult to specify the optimum intervention type in all cases. For this reason, as described above, historical data are used by the proposed system to determine the most suitable type of intervention per failure type.

4.5.1   Heuristic Approach.   For each failure type an intervention type must be determined and in the system the chosen type of intervention is then recommended for all sections of track which fail with that failure type.

An intervention type determination metric is applied to each failure type as follows;

```
foreach (run of data) {
    score the effectiveness of the following intervention
    using a metric M
}

foreach (intervention type) {
    calculate the average of all data runs which have a
    following intervention of that type
}

Select the intervention type with the highest average
    metric score
```

At face value, the algorithm appears simplistic yet effective, however the complexity lies within the production of a scoring metric (M), as an ineffective metric results in a poor choice of intervention type. A number of factors have been considered in the research reported herein. However, when considered in isolation, they were not found to be suitable and therefore some combination of factors had to be chosen as described below.

- Using the length of time to the next intervention work as the sole parameter is unsuitable since a metric based on this parameter would result in renewal being predominantly chosen. While renewal would result in the minimum amount of maintenance being performed over time it is not necessarily the most cost effective intervention.

- Using a metric solely based on the next failure type is also unsuitable. Intervention work which is applied to a section of track which does not cause the failure type to change from what it was before the intervention, may be regarded as remedying the symptoms rather than the underlying cause of failure (i.e. the intervention may be regarded as cosmetic). However, when track is deteriorating at an optimal rate (i.e. when there are no underlying faults, but rather the deterioration is a result of normal use alone) a metric based on the next failure type would tend to result in an intervention type being suggested that would cause the track to deteriorate in a non-optimal way (i.e. resulting in a failure being introduced to the track, or being unnecessarily treated).

The proposed system should be able to identify troublesome areas of railway track to the engineer. To this end, a procedure is proposed that enables the system to identify any sections of track where the selected intervention type's metric score is less than a defined threshold value. Further, using this process the system is able to give a list of the intervention types that have been applied in the past to any type of failure and determine their effectiveness at remedying the failures.

Conceptually, the intervention type determination process produces a directed non-fully connected graph of failure types, intervention types, and quality of intervention. Figure 8-25 shows a much simplified example.

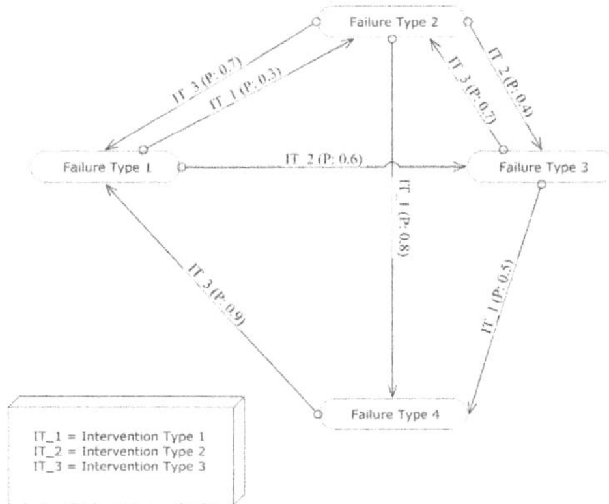

Figure 8-25.   Failure Type Linking

In practice it is more likely that for any given failure type and intervention type there will be more than one post intervention failure type. In such cases, a probability of the intervention resulting in the post intervention failure type is also included with the metric. This is based on the number of runs matching the pre and post failure types and the total number of runs with that intervention type for those failure types.

4.5.2   Life Cycle Costs.   The process of intervention type determination can be extended to take into account total life cycle costing. Existing solutions allow some total life cycle costing, however only on a macro scale (e.g. the costs associated with delaying a maintenance treatment by 6 months). By modifying the selection of the best average metric score in the intervention type determination metric to choose the intervention type with highest score of a second metric, a more sophisticated result can be obtained that takes into account factors such as cost and resource allocation. The second metric should take into account additional parameters such as the cost of the intervention work, age of components (to allow for older track to be replaced rather than repaired), and any other factors which may form part of the decision process normally used by an engineer.

5.   PROTOTYPE IMPLEMENTATION

In order to test the design of the proposed system, a prototype was developed as a proof of concept. In this section, the significant parts of the prototype are described and discussed.

5.1   System Training

Before any sections of track can be processed the system is initialised, this requires the user to carry out a number of processes, each of which are described in several dialog boxes (Figure 8-26). Since the engineer may not be fully conversant with the system, the dialog is designed to lead the engineer through the process of preparing the database for intervention planning; the following describes the function of each stage of this process.

1 Database connection: before any processing can be started, the location of the data is specified. Table 8-6 shows a partial knowledge base comprising geometry measurements taken over a section of line (AAV1100) from 29125km to 29625km, over a number of geometry recording runs (i.e. runs from 28/11/1997 to 17/02/2001).

2 Filtering for missing data: in the proposed system both geometry types and geometry records are removed if more than 60% of the geometry measurements or 60% of the geometry record is missing. During this phase a unique identifier (UID) is added to the geometry and work history values so that they can be referenced later. Since UID, Line Track ID, From, To, and Date are required they are not removed by filtering. However they are marked as linking columns so that they are not processed. Flag is removed since all values for this measurement are the same and therefore it would not be able to distinguish anything using this measurement. Measures of track geometry as specified by, Abs Vertical

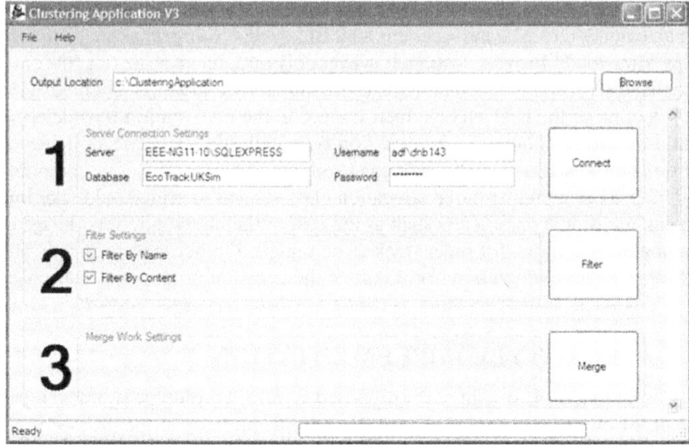

Figure 8-26.   System Preparation Dialog

Table 8-6.   Sample Geometry Data

LineTrack ID	Campaign ID	From	To	Date	Abs Ali	Abs Vert	Abs Twi	Abs Xlev	Abs Gaug
AAV1100	02/08/2000	29125	29250	02/08/2000	4.8	0	0	null	1445
AAV1100	05/06/1998	29125	29250	05/06/1998	4.5	0	0	null	1454
AAV1100	17/02/2001	29250	29375	17/02/2001	null	null	null	null	1445
AAV1100	05/06/1998	29250	29375	05/06/1998	5.1	0	1	null	1445
AAV1100	02/08/2000	29250	29375	02/08/2000	4.9	0	2	null	1441
AAV1100	04/11/1998	29250	29375	04/11/1998	null	null	null	null	1446
AAV1100	17/02/2001	29375	29500	17/02/2001	null	null	null	null	1441
AAV1100	28/11/1997	29375	29500	28/11/1997	2.4	0	0	null	1436
AAV1100	04/11/1998	29375	29500	04/11/1998	null	null	null	null	1439
AAV1100	02/08/2000	29375	29500	02/08/2000	2.5	0	0	null	1435
AAV1100	05/06/1998	29375	29500	05/06/1998	2.4	0	0	null	1440
AAV1100	28/11/1997	29500	29625	28/11/1997	null	null	null	null	null
AAV1100	17/02/2001	29500	29625	17/02/2001	3.5	0	0	null	1438
AAV1100	05/06/1998	29500	29625	05/06/1998	null	null	null	null	null

and Abs Gauge are kept as less than 60% of their values are missing. However, the measures of track geometry specified by Abs Twist and

Abs Crosslevel have over 60% of their values missing and are therefore discarded.

3 In order to produce sets of runs work data are merged together with the geometry data (i.e. a linked form using the unique identifiers previously added).

4 For each column of data null values are filled using the mean of values in the column. Whist this method has proved satisfactory so far, it is anticipated that further research will be performed to determine the effect of the various missing value substitution techniques described earlier (see Section 4.2.1).

5 Values are converted from absolute values to relative values in order to make a valid comparison between different track sections. A direct comparison is not possible as some geometry measurements may be different between track sections due to differences in track design. The process scales each run of data to an initial base value of 100, converting absolute values of geometry measurement to a percentage of the initial run measurement.

6 Clustering is performed. During this phase, an RPCL algorithm is applied to the last geometry measurement of each geometry run to determine failure types.

7 Cluster filtering is performed to remove excess clusters which do not correspond to failure types. A cluster is removed if it contains less than a specified number of geometry measurements (e.g. 20). This process removes clusters tending away from the geometry measurements (see Section 4.3.2).

8 Cluster definitions, defining the regions covered by each cluster, are produced.

9 Intervention levels and initial geometry levels are determined.

10 Work types and costs are determined.

11 Failure type linking is performed, resulting in a failure type linking chart (see Section 4.5.1).

5.2 Intervention Planning

Intervention planning can be performed once the system has been fully trained.

5.2.1   Selection Creation.   Before an intervention plan can be produced, the track sections to be analyzed are specified via the 'New Selection' dialog (Figure 8-27). The new selection dialog contains three parts; general settings allows naming and access rights to be set; the left hand tree list contains all possible track sections which can be used to form a selection; the right hand list specifies the track sections which have been added to the selection and provides a means (via the use of buttons) of adding and removing track sections from the selection. In Figure 8-27, a new selection named 'Selection 1' been created; the selection contains line track ID 00133 from 73234km to 74634km. Line track ID 01301 has been selected to add to the selection. In addition the user has used a feature to highlight all references containing 013 (this feature allows the quick selection of multi track sections conforming to a specific rule; once highlighted the sections can be quickly and easily added to a selection). Finally the user has specified that the system should open the selection once finalized (see Section 5.2.2 below).

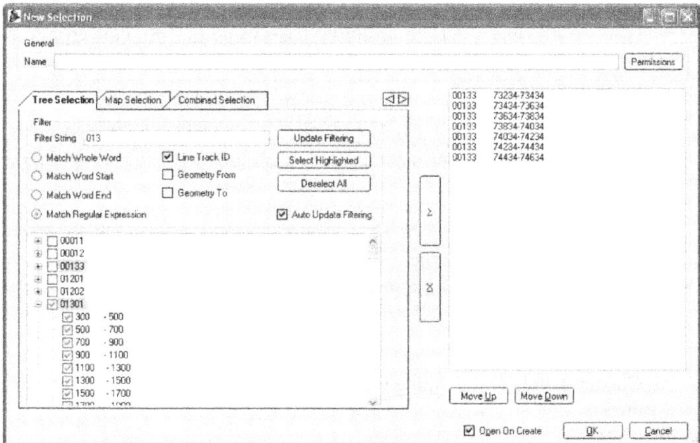

Figure 8-27.   Create Selection Dialog

5.2.2   Selection View.   The selection view (Figure 8-28) allows the user to display various pieces of information about the specified selection, including the most recent geometry measurements for each section of the selection. Geometry data can be viewed in a number of common formats and at various scales. In addition, historical data can be viewed by moving the mouse over the relevant section of track and geometry measurement. The current and historical failure type classifications can be displayed in a similar manner and are colour coded to match the settings dialog.

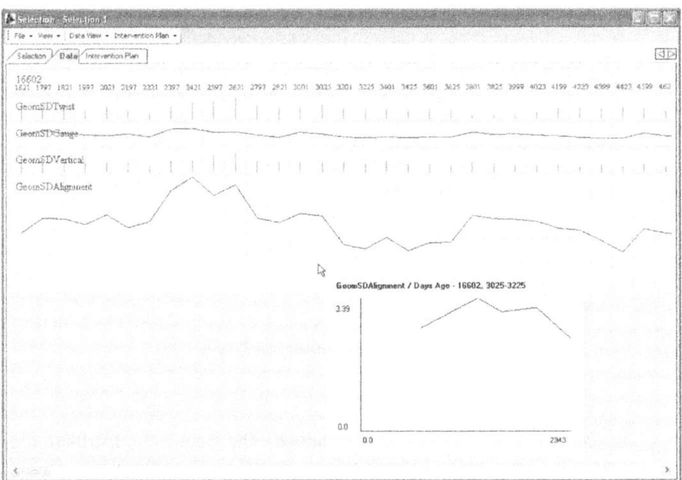

Figure 8-28.   Selection Data View

5.2.3   Intervention Plan.   The intervention plan view (Figure 8-29) is currently under development, and therefore only displays performed work history data, however in future versions of the prototype it is anticipated that it will also display recommended work. Each intervention is represented by a grey mark at the relevant section and date. Further information such as cost, location, and type can be displayed by moving the mouse over the corresponding intervention work item.

5.2.4   Analysis Tools.   To aid in the development and testing of the prototype, several tools were created as follows;

- A statistical analysis tool was developed which allows the analysis of the raw data. This tool allows the percentage and frequency distribution of missing values to be determined. For the former both the percentage of missing values per column (e.g. track geometry value) and by row (i.e.

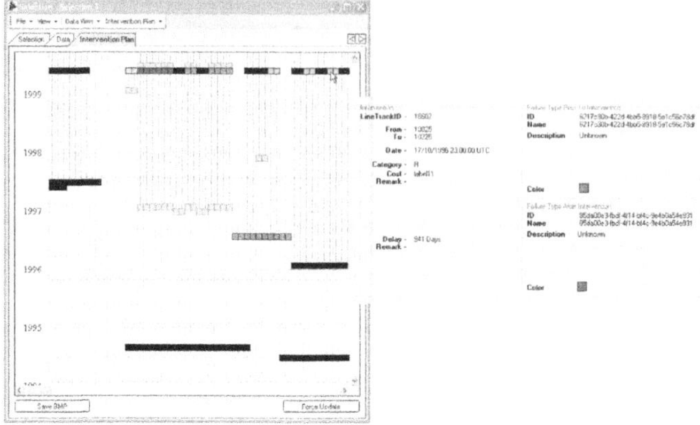

Figure 8-29.   Intervention Plan Dialog

by geometry record) can be determined. The frequency distribution of values specifies the top 15 most common values per column (as before). The statistical analysis tool can also be used to produce a raw view of each set of data (e.g. geometry and work history data as per figures 8-3 and 8-4). 8-7 shows a partial output for the statistic analysis tool. As can been seen, there is a large percentage of missing values indicating that the dataset is of poor quality.

- The Primary Classification Evaluation Tool was designed to display the effectiveness of the primary classification system. This is achieved by parsing the training set used in generating the primary classification system back through the primary classifier to compare the output of the classifier with the known failure type. The geometry data points are coloured depending on the result of the comparison. Green is used to indicate a positive match (i.e. the classifier has determined the correct failure type), aqua to indicate that the classifier determined a number of failure types of which one was correct, orange that multiple incorrect classifications were determined, and red to indicate the classification system produced one failure type which was incorrect. It was expected that beyond the region of initially good track quality, the clusters diverge and can be differentiated via the primary classification system. In the example presented in Figure 8-30, the classification rate is poor across

Table 8-7. Sample Statistical Data For Geometry Table

Column Name	Average	Number Of Missing Values	Number Of Missing Value Percentage
LineTrackID	NaN	0	0
CampaignID	NaN	0	0
GeomFrom	67064.53	0	0
GeomTo	67189.35	0	0
GeomDate	NaN	0	0
GeomFlag	NaN	0	0
GeomAbsAlignment	2.80	1270513	60
GeomAbsVertical	0.00	1272271	60
GeomAbsTwist	0.10	1272271	60
GeomAbsCrosslevel	0	1394180	66
GeomAbsGauge	1436.80	1282211	61
GeomAbsValue	2.77	282606	13
GeomSDAlignment	1.41	91456	4
GeomSDVertical	2.60	54429	2
GeomSDTwist	2.16	48440	2
GeomSDCrosslevel	1.74	1272140	60
GeomSDGauge	0	2091646	100
GeomSDValue	3.48	218747	10
GeomFaultsAlignment	0.02	1309606	62
GeomFaultsVertical	0.06	1274511	60
GeomFaultsTwist	0.08	1272828	60
GeomFaultsCrosslevel	0.00	1273976	60
GeomFaultsGauge	0.05	1297574	62
GeomFaultsValue	0	1349403	64
GeomSecurityAlignment	0.08	1316366	62
GeomSecurityVertical	2.89	1294169	61
GeomSecurityTwist	0.67	1272828	60
GeomSecurityCrosslevel	0.98	1291836	61
GeomSecurityGauge	0.24	1297474	62
GeomSecurityValue	0	1349403	64
GeomQuality	4.34	30208	1

the whole range of geometry values, indicating that the primary classification system is ineffective. For the older track geometry recordings (i.e. rightmost), the classification appears to be somewhat improved, however this is most likely to be due to overlapping clusters becoming increasingly large, until the largest just includes all of the data.

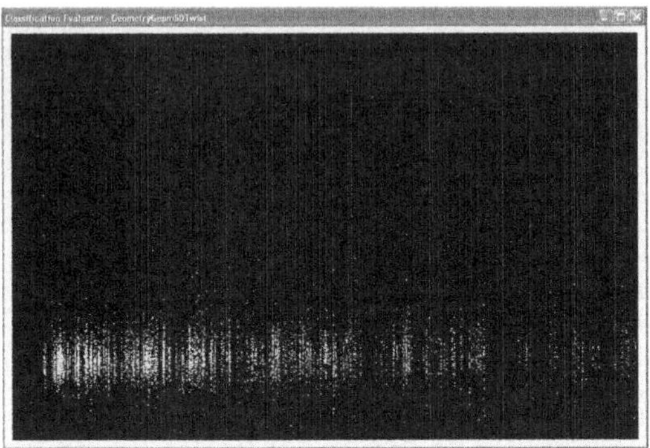

Figure 8-30. Primary Classification Evaluation Tool

- Failure types can be analyzed via the Failure Type Linking Analysis Tool. The tool allows each failure type to be viewed and allows the effect of maintenance to be determined. In the example shown in Figure 8-31, a track section which deteriorates with Failure Type 1 has been treated with maintenance type T and then subsequently deteriorates with Failure Type 3. In some cases, as can be seen with Failure Type 3, given any type of further maintenance, the track section will still deteriorate with Failure Type 3. Such a scenario may be due to either the maintenance types performed on this failure type in the past having only treated the symptoms (and hence the underlying cause remains), or the track is deteriorating in the optimal manner (i.e. there are no specific faults with the track section but rather it is failing due to general use). In Figure 8-31, the pop-up box shows information corresponding to the link between Failure Type 3 and Failure Type 2. In addition information relevant to the pre and post intervention is displayed, along with information relevant to the type of intervention. In cases where given a particular failure type and intervention type, there are multiple post intervention failure types, the probability of post intervention failure type is displayed on the link. The system displays the best probability / link in the given example for clarity of presentation.

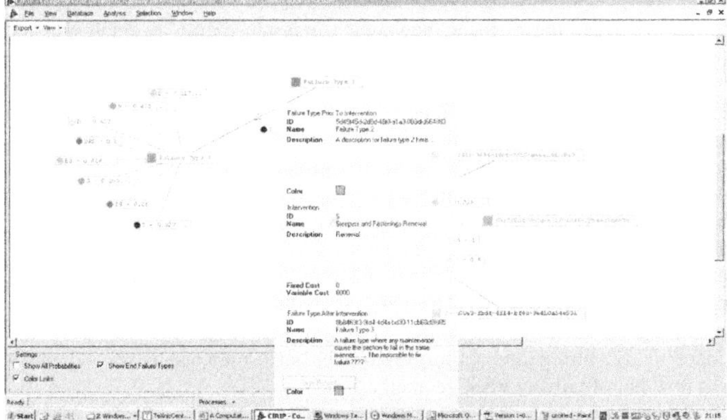

Figure 8-31. Failure Type Linking Analysis Tool

6. CONCLUSIONS

Existing decision support systems for railway track intervention planning are sub-optimal in dealing with the complexity of intervention planning. In particular, the fact base containing the rules in a typical expert system structure of existing decision support tools may be regarded as having overly simplistic deterioration models which are unable to describe the complex interactions between track components.

To address these issues a data driven computational intelligence approach to railway track intervention planning system was presented in this chapter. The proposed system uses a variety of computational intelligence techniques, including clustering to determine the various failure types applicable to track deterioration, evolutionary algorithms to produce deterioration models for each failure type, and a heuristic based approach to determine the most appropriate maintenance to perform per failure type.

In the proposed system, a number of novel processes and techniques have been adopted, including the methodology of deriving the functionality of the system from the available data itself without the need for expert judgment. Furthermore, the proposed system can determine the possible types of failure by which a selected section of track can deteriorate (including the degree of track affected by such failure), and the interaction between intervention types and failure types. While the system may not be able to provide names for track failure types in terms readily recognised by a railway engineer, combined with the knowledge of the engineer, the system can produce a highly detailed history (or prediction) of the changing state of a given section of track.

From a computer science perspective, the system also contains novel aspects, e.g., techniques used to determine failure types and the production of deterioration models. In the determination of failure types, a clustering algorithm was used on railway track geometry data. To overcome the typically used trial and error approach to tune clustering algorithms to a training set, a known number of failure types and members was produced which mimicked the behaviour of a simplistic real data set. The training set was processed using the clustering algorithm until the output from the clustering algorithm matched the known properties of the training set. The two stage evolutionary algorithm used for the production of deterioration models is also innovative. This system, once tuned, enables curve fitting (and optimization) to be performed on any set of data, and not just that of railway track geometry data.

6.1   Strengths of the Proposed System

- Improvement over time: The proposed system can be retrained as and when more data are available, thus improving the accuracy of the output. As more track condition data becomes available the system may find new undiscovered failure types. As a result, more data will become available for the production of deterioration models, and also for use in the heuristic based approach to intervention type determination.

- The behaviour of the system is defined by historical data: The behaviour of existing systems is set by engineers, and therefore any mistakes or omissions added to the fact base results in a system that performs incorrectly, or at best non-optimally. Using behaviour generated from historical data ensures that the system will perform at worst, as well as systems used in the past.

- Identification of weak areas of diagnostic: The existing systems give only very general indications of the quality of intervention plans, whereas the proposed system is able to identify problem areas more specifically.

6.2   Weaknesses of the Proposed System

The weaknesses of the proposed system are related to the reliance of the system on historical data;

- Since the proposed system is defined by historical data, a high quality dataset with few errors and missing values must be used to train the system.

- The historical data used for training the proposed system must already contain railway track which has been treated in a suitable manner, otherwise the system will not initially perform well. Furthermore, the quality of the intervention plans produced is dependent on the quality of the treatment within the historical data. While the system is able to identify poorly treated sections of track historically, it is not able to determine the effect of a treatment previously unused upon a particular failure type.

- The historical data must be representative of the types of track for which intervention plans are being generated. If a track section has a failure type previously unseen by the system, then the output of the system is undefined and unknown.

6.3   Areas for Further Research

The following items of further research are recommended;

- Uncertainty of classification of sections with failure types unknown to the system (i.e. not in historical data)

- Varying section size

- Total life cycle costing; optimization; including coherence processing.

References

Association of Train Operating Companies (2006), *Ten-Year European Rail Growth Trends*, The Association of Train Operating Companies, London, UK., 1 July 2006

Burrow, M.P.N., Ghataora, G.S & Bowness, D. (2004), Analytical track substructure design. *ISSMGE TC3 International Seminar on Geotechnics in Pavement and Railway Design and Construction*, NTUA - Athens, 16-17 December 2004., pp. 209–216

Cheung, Y. (2004), *A Competitive and Cooperative Learning Approach to Robust Data Clustering*, IASTED International Coference on Neural Networks and Computational Intelligence, 23-25 February 2004, pp. 131–136

Cope, G.H.(1993), *British Railway Track — Design, Construction and Maintenance*, The Permanent Way Institution, Eco Press. Loughborough, UK., 1993

Coulouris, G.F., Dollimore, J. & Kindberg, T. (2000), *Distributed Systems: Concepts and Design*, Third Edition. International Computer Science, 23 August 2000, pp. 28–64

ERRI (1994), *EcoTrack - Decision Support System for Permanent Way Maintenance and Renewal - Specifications 1. General Concept*, ERRI, April 1994

Esveld, C. (2001), *Modern Railway Track*, Second Edition. MRT-Productions, Zaltbommel, The Netherlands., August 2001

Jovanovic, S. (2000), *Optimal Resource Allocation Within the Field of Railway Track Maintenance and Renewal*, Railway Engineering 2000, London, 5-6 July 2000.

Jovanovic, S. & Esveld, C. (2001), *An Objective Condition-based Decision Support System for Long-term Track Maintenance and Renewal Planning*, 7th International Heavy Haul Conference, Australia, 10-14 June 2001, pp. 199–207

Jovanovic, S. & Zaalberg, H. (2000), *EcoTrack - Two Years of Experience*, Rail International, April 2000, pp. 2–8

King, I. & Lau, T. (1999), *Non-Hierarchical Clustering with Rival Penalized Competitive Learning for Information Retrieval*, First International Workshop on Machine Learning and Data Mining in Pattern Recognition, 16-18 September 1999, pp. 116-130

Leeuwen, R.V. (1996), *The EcoTrack Project - Effective Management of Track Maintenance*, Belgium National Railway Company, Brussels, 1996, pp. 1–16

Moore, A.W. (2004), *K-means and Hierarchical Clustering*, Carnegie Mellon University, 8 October 2004.

Network Rail (2006), *Business Plan*, Network Rail, UK., 4 April 2006

Stirling, A.B., Roberts, C.M., Chan, A.H.C., Madelin, K.B. & Bocking, A. (1999), *Development of a Rule Base (Code of Practice) for the Maintenance of Plain Line Track in the UK to be used in an Expert System*, Railway Engineering 1999, London, 26-27 May 1999

Stirling, A.B., Roberts, C.M., Chan, A.H.C., Madelin, K.B. & Vernon, K. (2000), *Trial of an Expert System for the Maintenance of Plain Line Track in the UK*, Railway Engineering 2000, London, 5-6 July 2000

Stirling, A.B., Roberts, C.M., Chan, A.H.C. & Madelin, K.B. (2000), *Prototype Expert System for the Maintenance and Renewal of Railway Track*, Freight Vehicle Design Workshop, Manchester, 2000

Rivier, R. & Korpanec, I. (1997), *EcoTrack - A Tool to Reduce the Life Cycle Costs of the Track*, World Congress on Railway Research, 16-19 November 1997, pp. 289–295

Rivier, R.E. (1998), *EcoTrack - A Tool for Track Maintenance and Renewal Managers*, Computational Mechanics Publications, Institute of Transportation and Planning, Swiss Federal, September 1998, pp. 733–742

Roberts, C. (2001), *A Decision Support System for Effective Track Maintenance and Renewal, PhD Thesis*, The University of Birmingham, January 2001

Roberts, C. (2001), *Decision Support System for Track Renewal Strategies and Maintenance Optimization*, Railway Engineering 2001, 30 April-1 May 2001

Zaa, P.H. (1998), *Economizing Track Renewal and Maintenance with EcoTrack*, Cost Effectiveness and Safety Aspects of Railway Track, Paris, 1998

Chapter 9

A CO-EVOLUTIONARY FUZZY SYSTEM
FOR RESERVOIR WELL LOGS
INTERPRETATION

Tina Yu[1] and Dave Wilkinson[2]

[1]*Memorial University of Newfoundland, St. John's, NL A1B 3X5, Canada,* [2]*Chevron Energy Technology Company, San Ramon, CA 94583, USA*

Abstract   Well log data are routinely used for stratigraphic interpretation of the earth's subsurface. This paper investigates using a co-evolutionary fuzzy system to generate a well log interpreter that can automatically process well log data and interpret reservoir permeability. The methodology consists of 3 steps: 1) transform well log data into fuzzy symbols which maintain the character of the original log curves; 2) apply a co-evolutionary fuzzy system to generate a fuzzy rule set that classifies permeability ranges; 3) use the fuzzy rule set to interpret well logs and infer the permeability ranges. We present the developed techniques and test them on well log data collected from oil fields in offshore West Africa. The generated fuzzy rules give sensible interpretation. This result is encouraging in two respects. It indicates that the developed well log transformation method preserves the information required for reservoir properties interpretation. It also suggests that the developed co-evolutionary fuzzy system can be applied to generate well log interpreters for other reservoir properties, such as lithology.

Keywords: reservoir modeling and characterization, fuzzy logic, co-operative co-evolution, time series, well logs interpretation, genetic programming.

1.   INTRODUCTION

In reservoir characterization, well log data are frequently used to interpret physical rock properties such as lithology, porosity, pore geometry, depositional facies and permeability. These properties are keys to the understanding of an oil reservoir and can help determining hydrocarbon reserves and reservoir producibility. Based on the information, decisions of where to complete a well, how to stimulate a field, and where to drill next, can be made to maximize profit and minimize risk.

T. Yu and D. Wilkinson: *A Co-Evolutionary Fuzzy System for Reservoir Well Logs Interpretation*, Studies in Computational Intelligence (SCI) **88**, 199–218 (2008)

www.springerlink.com

Well log data, ranging from conventional logs, such as spontaneous potential, gamma ray, and resistivity, to more advanced logging technology, such as Nuclear Magnetic Resonance (NMR) logs, are sequence of curves indicating the properties of layers within the earth's subsurface. Figure 9-1 gives an example of gamma ray, neutron and spontaneous potential (SP) logs. The interpreted lithology is listed on the left-hand side.

Well log interpretation is a time-consuming process, since many different types of logs from many different wells need to be processed simultaneously. This paper investigates using a co-evolutionary fuzzy system to generate a well-log interpreter that can process well log data and interpret reservoir permeability automatically. The developed methodology has 3 steps:

- Transform well log data into fuzzy symbols which maintain the character of the original log curves.

- Apply a co-evolutionary fuzzy system to generate a fuzzy rule set that classifies permeability ranges.

- Use the fuzzy rule set to interpret well logs and infer the permeability ranges.

Similar to time series, well logs are sequential data, which are indexed by the depth under earth's surface where the data were collected. To interpret earth properties, similar consecutive log data can be grouped into blocks, since rock properties formation is frequently developed in layers. By examining blocked wells logs across the same depth, geologists are able to detect earth properties at that particular layer.

In this research, we developed a computer system to carry out the well log blocking process. Additionally, the numerical data are transformed into fuzzy symbols (Yu and Wilkinson, 2007). Fuzzy symbol representation has advantages over its numerical counter-part in that it is easier for computers to manipulate and to carry out the interpretation task. Meanwhile, because fuzzy symbols have no precise boundaries, they allows efficient interpretation under the uncertainty embedded in thc data sets.

The second step of the process uses a co-evolutionary fuzzy system to extract fuzzy rule patterns in the transformed well logs fuzzy symbols to distinguish different permeability ranges. Since permeability can be divided into more than one ranges (3 in this study), the evolutionary system maintains multiple populations, each of which evolves rules that classify one permeability range from others. These populations co-evolve to produce a combined fuzzy rule set that can classify all possible permeability ranges. Once completed, this fuzzy rule set can be used to interpret permeability of other wells with similar geological characteristics.

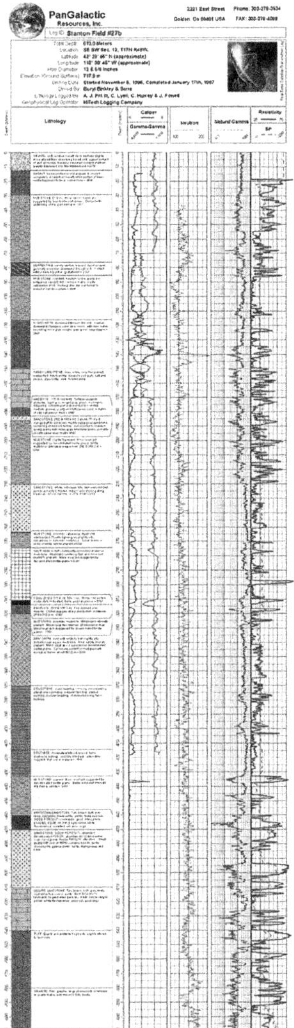

Figure 9-1. An example of gamma ray, neutron and spontaneous potential logs. The interpreted lithology is listed on the left-hand side.

We have tested the developed method on well log data collected from oil fields in offshore West Africa and the results are very encouraging. Based on this initial study, we are currently applying the system to develop a reservoir lithology interpreter, which requires a more sophisticated co-evolutionary model to interpret 5 different types of lithology. In this case, 4 populations are co-evolved together to accomplish this task.

We organize the paper as follows. Section 2 presents the methodology to transform well log data into fuzzy symbols. Information about the testing well log data and the transformed results are given in Section 3. Section 4 introduces the co-evolutionary fuzzy system developed to generate fuzzy rules. After that, the experimental setup for fuzzy rule generation is given in Section 5. In Section 6, we report the experimental results. Analysis and discussion are then provided in Section 7. Finally, Section 8 concludes the paper.

2. WELL LOG TRANSFORMATION

The fuzzy symbolic representation is an approximation of well logs data that maintains the trend in the original data. The transformation process has four steps: 1) segmentation of the numerical well log data; 2) determining the number of segments; 3) symbol assignment; and 4) symbol fuzzification. These steps are explained in the following sub-sections.

2.1 Segmentation

Well log segmentation involves partitioning log data into segments and using the mean value of the data points falling within the segment to represent the original data. In order to accurately represent the original data, each segment is allowed to have arbitrary length. In this way, areas where data points have low variation will be represented by a single segment while areas where data points have high variation will have many segments.

The segmentation process starts by having one data point in each segment. That is the number of segments is the same as the number of original data points. Step-by-step, neighboring segments (data points) are gradually combined to reduce the number of segments. This process stops when the number of segments reaches the predetermined number.

At each step, the segments whose merging will lead to the least increase in *error* are combined. The *error* of each segment is defined as:

$$error_a = \sum_{i=1}^{n} (d_i - \mu_a)^2$$

where n is the number of data points in segment a, μ_a is the mean of segment a, d_i is the ith data point in segment a.

This approach is similar to the Adaptive Piecewise Constant Approximation proposed by (Keogh et al., 2001) and SAX (Lin et al., 2003). However, our

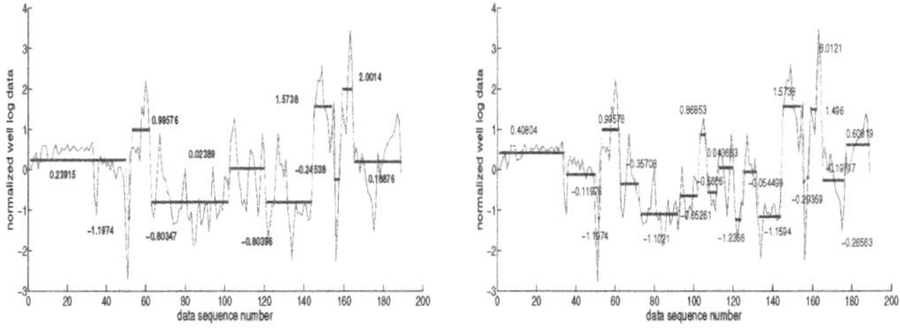

Figure 9-2.   10 segments.   *Figure 9-3.*   20 segments.

method has an extra component that dynamically determines the number of segments (see Section 2.2). Another similar work using a different approach to determine the number of segments is reported in (Abonyi et al., 2005).

Figure 9-2 is an example of a well log with 189 data points, which are partitioned into 10 segments. The same data are partitioned into 20 segments in Figure 9-3. The average value of the data points within each segment is used to represent the original data.

2.2   Number of Segments

Although a larger number of segments capture the data trend better, it is also more difficult to interpret. Ideally, we want to use the smallest possible number of segments to capture the trend of the log data. Unfortunately, these two objectives are in conflict: the total *error* of all segments monotonically increases as the number of segments decreases (see Figure 9-4). We therefore devised a compromised solution where a penalty is paid for increasing the number of segments. The new *error* criterion is now defined as the previous total *error plus* the number of segments:

$$f = N + \sum_{i=1}^{N} error_i \quad \text{where } N \text{ is the number of segment.}$$

During the segmentation process, the above f function is evaluated at each step when 2 segments were combined. As long as this value f is decreasing, the system continues to merge segments. Once f starts increasing, it indicates that farther reducing the number of segments will sacrifice log character, hence the segmentation process terminates. For the 189 data points in Figure 9-2, the final number of segments is 50 (see Figure 9-5).

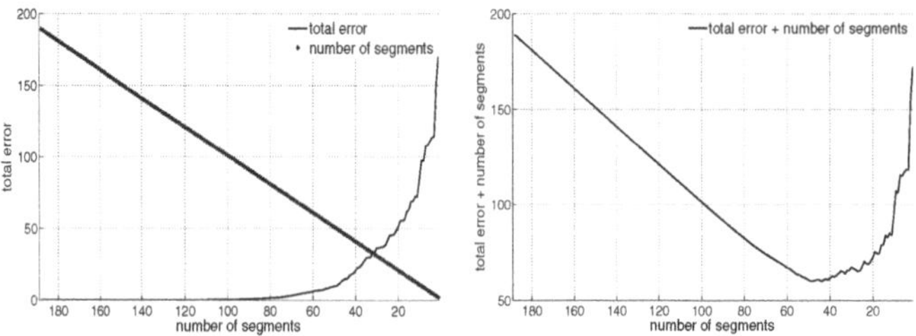

Figure 9-4. number of segments vs. total error.

Figure 9-5. a compromised solution.

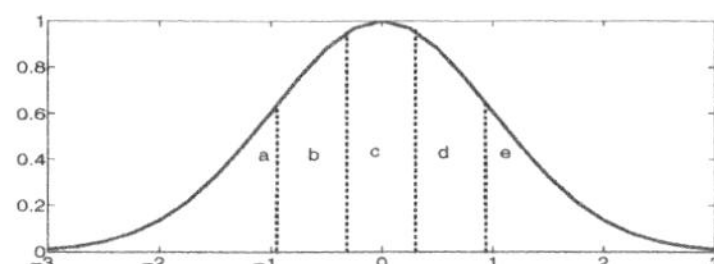

Figure 9-6. Using 4 breakpoints to produce 5 symbols with equal probability.

2.3   Symbol Assignment

Segmented well logs are represented as a set of numerical values, $\overline{WL} = \overline{s_1}, \overline{s_2}, \overline{s_3} \ldots$, where $\overline{s_i}$ is the mean value of the data within the ith segment. This numerical representation is farther simplified using symbols. Unlike numerical values, which are continuous, symbols are discrete and bounded. This makes it easy for any subsequent computer interpretation scheme.

While converting the numerical values into symbols, it is desirable to produce symbols with equal-probability (Apostolico et al., 2002). This is easily achieved since normalized sequence data have a Gaussian distribution (Larsen and Marx, 1986). We therefore applied z-transform to normalize the data and then determined the breakpoints that would produce n equal-sized areas under the Gaussian curve, where n is the number of symbols. Figure 9-6 gives the four breakpoints -0.84, -0.25, 0.25 and 0.84 that produce 5 symbols, a, b, c, d, e, with equal probability. If only 3 symbols (a, b and c) are used, the breakpoints are -0.43 and 0.43.

Once the number of symbols, hence the breakpoints have been decided, we assign symbols to each segment of the well logs in the following manner: All segments have mean values that are below the smallest breakpoint are mapped to the symbol a; all segments have mean values that are greater than or equal to the smallest breakpoint and less than the second smallest breakpoint are mapped

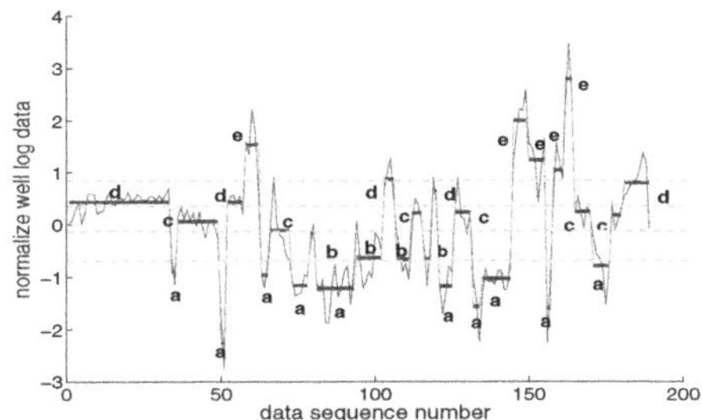

Figure 9-7.   A well log transformed using 5 symbols.

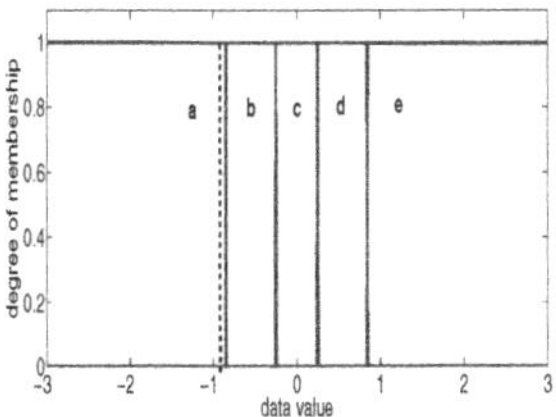

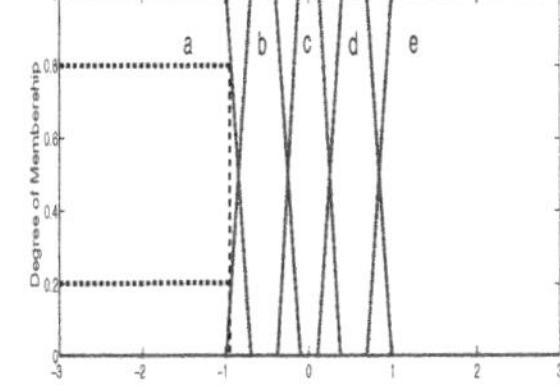

Figure 9-8.   A segment with mean -0.9 is transformed as a crisp symbol a.

Figure 9-9.   A segment with mean -0.9 is transformed as fuzzy symbol a (80%) and b (20%).

to the symbol b and so on. Figure 9-7 gives a well log that is transformed using 5 symbols.

2.4   Symbol Fuzzification

While some segments are clearly within the boundary of a particular symbol region, others may not have such clear cut. For example, in Figure 9-7, there are 3 segments that lie on the borderline of regions a and b. A crisp symbol, either a or b, does not represent its true value. In contrast, fuzzy symbols use membership function to express the segment can be interpreted as symbol a and b with some possibility. As an example, with the crisp symbol approach, a segment with mean -0.9 is assigned with symbol a with 100% possibility (see Figure 9-8). Using fuzzy symbols designed by trapezoidal-shaped membership functions, the segment is assigned with symbol a with 80% possibility and symbol b with 20% possibility (see Figure 9-9). Fuzzy symbol representation is more expressive in this case.

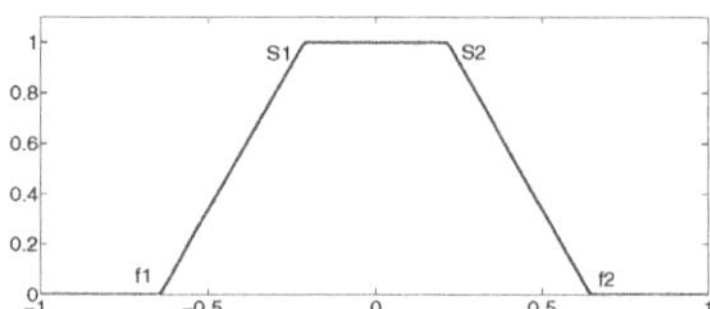

Figure 9-10. The 4 parameters, f1, f2, s1, s2, that define a trapezoidal-shaped membership function.

In fuzzy logic, a membership function (MF) defines how each point in the input space is mapped into a membership value (or degree of membership) between 0 and 1. The input space consists of all possible input values. In our case, z-normalized well log data have open-ended boundaries with mean 0. When 5 symbols are used to represent a well-log, 5 membership functions are defined, one for each of the 5 symbols.

To design a trapezoidal-shaped membership function, 4 parameters are required: f_1 and f_2 are used to locate the 'feet' of the trapezoid and s_1 and s_2 are used to locate the 'shoulders' (see Figure 9-10). These four parameters are designed in the following way.

Let c_1 and c_2 be the breakpoints that define a symbol x and $c_2 > c_1$:

$$d = \frac{c_2 - c_1}{4}$$
$$f_1 = c_1 - d;\, s_1 = c_1 + d;\, s_2 = c_2 - d;\, f_2 = c_2 + d$$

There are two exceptions: symbol a has $f_1 = c_1$ and symbol e has $f_2 = c_2$. Table 9-1 gives the four parameters used to design the membership functions for each symbol.

Once the 4 parameters are decided, the membership function f is defined as follows:

$$f(x, f_1, f_2, s_1, s_2) = \begin{cases} 0 & \text{if } x \leq f_1, \\ \frac{x - f_1}{s_1 - f_1} & \text{if } f_1 \leq x \leq s_1. \\ 1 & \text{if } s_1 \leq x \leq s_2, \\ \frac{f_2 - x}{f_2 - s_2} & \text{if } s_2 \leq x \leq f_2. \\ 0 & \text{if } f_2 \leq x, \end{cases}$$

Using the described fuzzy symbol scheme, the 10 segments lying between the two symbol regions in Figure 9-7 were mapped into fuzzy symbols shown in Figure 9-11.

Table 9-1.   Parameters used to design the trapezoidal-shaped membership function for each symbol.

data	symbol	f_1	s_1	s_2	f_2
well-log	a	-3	-3	-0.9875	-0.6925
	b	-0.9875	-0.6925	-0.3975	-0.1025
	c	-0.375	-0.125	0.125	0.375
	d	0.1025	0.3975	0.6925	0.9875
	e	0.6925	0.9875	3	3
perm	a	-3	-3	-0.645	-0.215
	b	-0.645	-0.215	0.215	0.645
	c	0.215	0.645	3	3

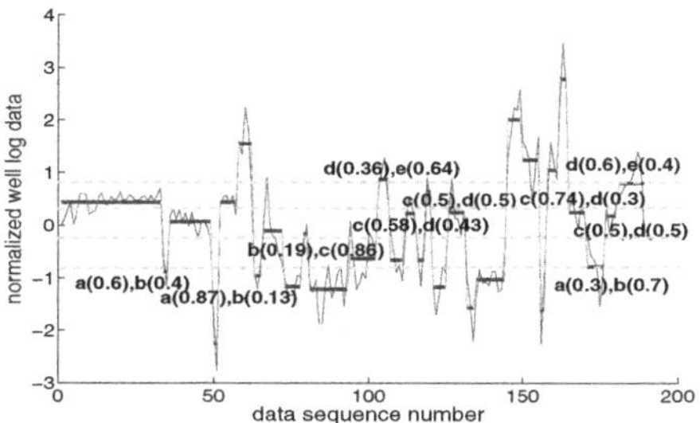

Figure 9-11.   A well log represented with fuzzy symbols.

In most cases, a reservoir well has multiple logs. To carry out the described transformation process, a reference log is first selected for segmentation. The result is then used to segment the other logs in the same well. After that, fuzzy symbols are assigned to each segmented data.

3.   WELL LOG DATA

We tested the developed transformation method on 2 sets of well log data collected from an offshore West Africa field. The first set is from Well A and contains 227 data points while the second set is from Well B and contains 113 data points. Each well has 3 different logs: PHI (porosity), $RhoB$ (density) and DT (sonic log). Additionally, V-shale (Volume of shale) information has been calculated previously (Yu et al., 2003). The core permeability data are available and will be used to test the evolved fuzzy rules.

Since permeability is the interpreted target, it is chosen as the reference log to perform segmentation described in Sections 2.1 and 2.2. For symbol assignment, permeability has 3 possible symbols, a, b and c, representing *low*, *medium* and *high* permeability. The 3 well logs and V-shale, however, have 5 possible symbols, a, b, c, d, e. This allows the evolved fuzzy rules to have a finer granularity in interpreting well log data. Figures 9-12, 9-13, 9-14, 9-15 and 9-16 give the transformed logs in Well A. The resulting transformations give sensible blocking and resemble the original log curves reasonably well. Due to space constraint, the results of Well B, which have a similar pattern, are not shown here.

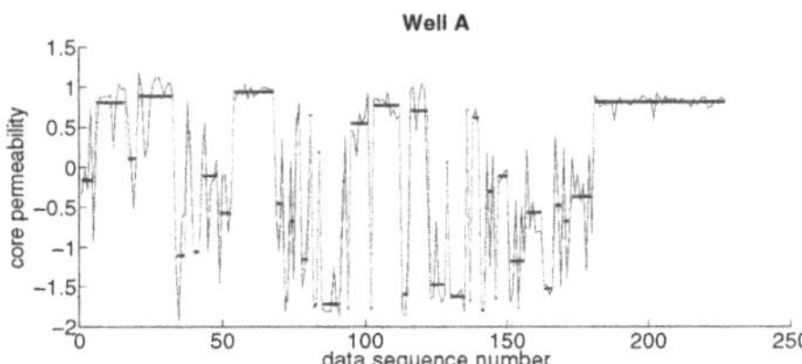

Figure 9-12. The transformed core permeability (k).

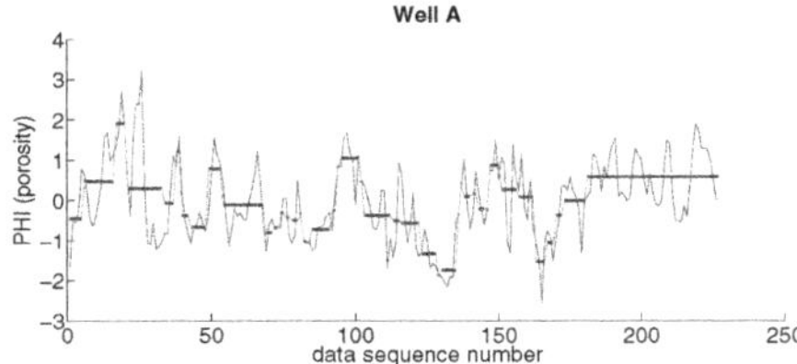

Figure 9-13. The transformed PHI log.

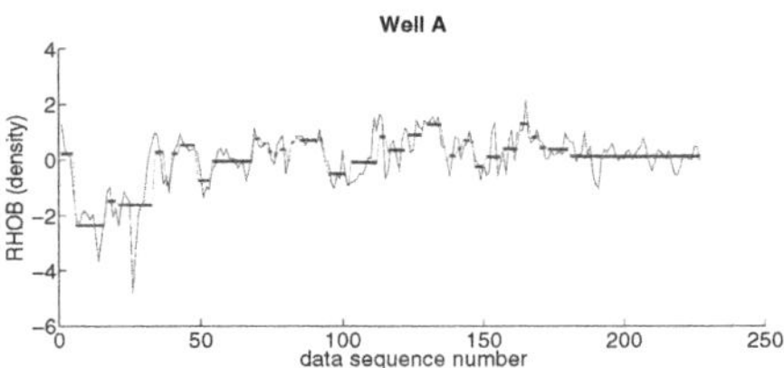

Figure 9-14. The transformed RHOB log.

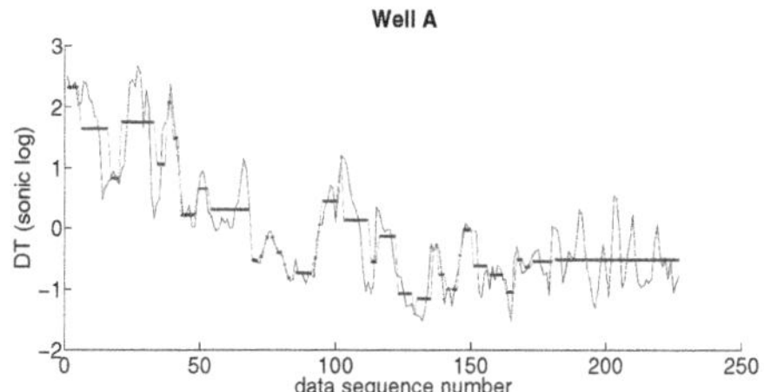

Figure 9-15.   The transformed DT log.

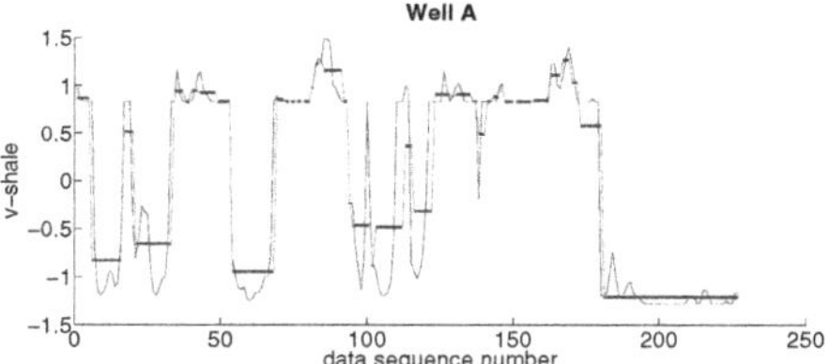

Figure 9-16.   The transformed V-shale data.

After the transformation process, all logs in Well A have 43 segments and all logs in Well B have 15 segments. Among the 43 permeability segmentations in Well A, 22 are low-permeability (symbol a), 9 are medium-permeability (symbol b) and 12 are high-permeability (symbol c). The number of low, medium and high permeability segments in Well B is 6, 2 and 7 respectively.

4.   CO-EVOLUTIONARY FUZZY SYSTEM

Using the transformed well log data, we applied a co-evolutionary fuzzy system to identify rule patterns that can interpret well logs having high, medium or low permeability. The interpretation task is decomposed into two sub-tasks: the first one separates one permeability range data from the rest of the data and the second one distinguishes another permeability range data from the others. By combining the two sub-solutions using an if-then-else construct, the final solution is able to determine whether a well log segment has either high, medium or low permeability.

We adopted a co-operative co-evolution approach to address these two sub-problems (Potter and Jong, 1994; Potter and Jong, 2000). In this approach, two populations are maintained, each of which is evolved toward one of the two

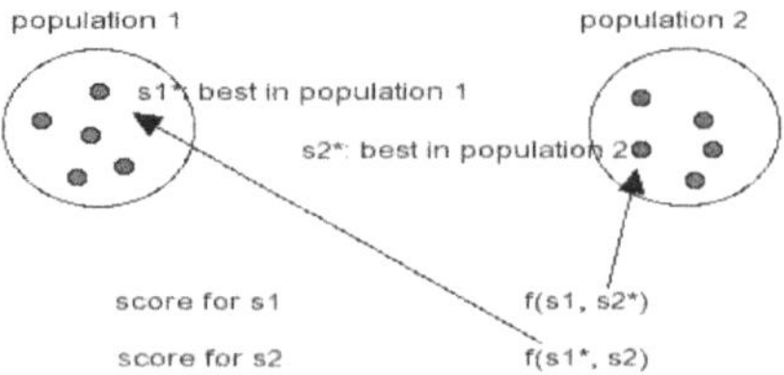

Figure 9-17.　　The co-operative co-evolution model.

different sub-goals. However, to encourage their co-operation to evolve the best overall permeability interpreter, the fitness of an evolved rule is determined by how well it collaborates with the rules evolved in the other population. In terms of implementation, a rule from one population is combined with the *best* rule in the other population and the performance of this combined rule-set defines the fitness of the rule in the current population. Figure 9-17 illustrates the described co-evolution mechanism.

There are other works using this co-evolutionary model to evolve fuzzy rules. For example, fuzzy co-co (Pena-Reyes and Sipper, 2001) maintains two populations: one evolves membership functions and the other evolves fuzzy rules.

4.1　　Fuzzy Rule Generation

The co-evolutionary system is implemented in a genetic programming(GP) system called PolyGP (Yu, 2001), which has a type system to perform type checking during rules evolution. In this way, the evolved rules (genotype and phenotype) are always type checked prior to fitness evaluation. There are other methods to evolve type-correct solutions (Yu and Bentley, 1998). For example, (Bentley, 2000) mapped type-incorrect fuzzy rules to correct ones using a repair method.

Table 9-2 gives the functions and terminals with their type signatures for the GP system to evolve type-correct fuzzy rules. The 3 well logs (*PHI,RHOB, DT*) and v-shale have a vector type of 5 values, each of which specifies the degree of membership to the 5 symbols a, b, c, d and e. For example, a segment with mean value 0.9 has a vector values [0.8, 0.2, 0, 0, 0]. The function *is-a, is-b, is-c, is-d* and *is-e* take a vector as argument and returns the degree of membership belongs to symbol a, b, c, d and e respectively. For example, *is-a*[0.8, 0.2, 0, 0, 0] = 0.8. Three fuzzy operators used to construct fuzzy rules are *and, or* and *not*: $and(x, y) = min(x, y)$, $or(x, y) = max(x, y)$, $not(x) = 1 - x$. Figure 9-18 gives an evolved fuzzy rule example.

Table 9-2.   Function and Terminal Sets

Function Terminal	Type
is-a	[float,float,float,float,float]→float
is-b	[float,float,float,float,float]→float
is-c	[float,float,float,float,float]→float
is-d	[float,float,float,float,float]→float
is-e	[float,float,float,float,float]→float
and	float→float→float
not	float→float
or	float→float→float
PHI	[float,float,float,float,float]
RHOB	[float,float,float,float,float]
DT	[float,float,float,float,float]
v-shale	[float,float,float,float,float]

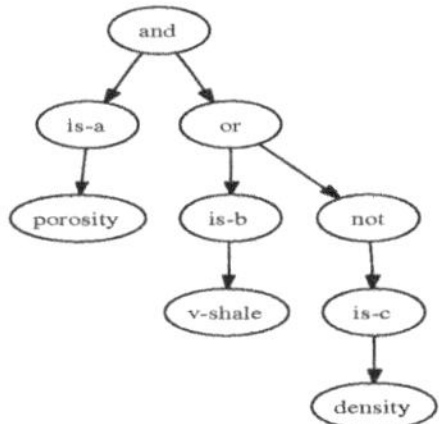

Figure 9-18.   An evolved fuzzy rule example.

To work with this fuzzy rule tree representation, we employed four genetic operators in this study: homologous crossover, *and*-crossover, *or*-crossover and mutation. Homologous crossover selects common location in both parent trees to carry out the crossover operation. The *and*-crossover combines two parent rules into one rule using the *and* operator. The *or*-crossover combines two parent rules into one rule using the *or* operator. The mutation operation can perform sub-tree, function and terminal mutations, depending on the selected mutation location.

4.2   Fuzzy Rules Evaluation

After evaluation, a fuzzy rule produces a numerical value between 0 and 1. This value indicates the degree of membership the data belongs to the classified permeability. We uses a simple defuzzification mechanism to interpret the result:

if the degree of membership is greater than or equal to 0.5, the data belongs to the classified range.

To assign a fitness to the evaluated fuzzy rule, the rule is first combined with the *best* rule in the other population using the following template:

$$if\ rule\text{-}1 \geq 0.5$$
$$then\ high\text{-}permeability$$
$$else\ if\ rule\text{-}2 \geq 0.5$$
$$then\ low\text{-}permeability$$
$$else\ medium\text{-}permeability.$$

where *rule-1* is a rule from the first population and *rule-2* is a rule from the second population. If the evaluated rule is from the first population, the *best* rule from the second population is used to complete the template. If the evaluated rule is from the second population, the *best* rule from the first population is used to complete the template. This combined if-then-else rule is then tested on the training data and the interpretation results are compared with the transformed permeability. If the if-then-else rule gives the correct interpretation, it is a hit. The percentage of the hit among the training data is the fitness of the evaluated rule. To promote shorter and more readable rules to be evolved, rules with length more than 100 nodes are penalized. Also, the *best* rule in each population is updated at the beginning of every generation, so that a good rule can be immediately used to combine with rules in the other population and impact evolution.

5.   EXPERIMENTAL SETUP

Both Well A and B have a greater number of high and low permeability data segments than medium-permeability data segments. We therefore used one population to evolve rules that separate high-permeability data segments and the other population to evolve rules that distinguish low-permeability data segments. In this way, both populations have a balanced number of positive and negative samples, which is important to train robust classifiers.

We used Well A data to train the fuzzy rules. The final best rule was then tested on Well B data. The crossover rates used are as follows: 20% for homologous crossover, 10% for *and*-crossover and 10% for *or*-crossover. Mutation rate is 50%. When no genetic operation is executed, an identical copy of one parent is copied over to the next generation.

The selection scheme is a tournament with size 2. We set the population size as 100 to run for 1,000 generations, where at each generation, the population is 100% replaced by the offspring except one copy of the elite (the best) which is kept and carried over to the new generation.

By combining two rules from two populations, it is sufficient to classify three permeability ranges. However, the order of their combination can effect the classification accuracy. This is because *rule-1* is evaluated first, according to the rule template. Once *rule-1* makes a wrong interpretation, *rule-2* can not correct it. Consequently, *rule-1* has a stronger impact than *rule-2* on the performance of the combined rule.

To achieve a better interpretation accuracy, it is desirable to have the rule which has better accuracy be *rule-1*. Unfortunately, we do not know in advance which of the two rules will give better accuracy. We therefore made two sets of experimental runs. In the first set, *rule-1* is the rule that identifies high-permeability data segments. In the second set, *rule-1* is the rule that classifies low-permeability data segments. Fifty runs were made for each of the two setups and their final best rules were collected for evaluation.

6.   RESULTS

Figure 9-19 gives the results of the two sets of runs. As shown, the fuzzy rules which classify high-permeability first produce better results: the average fitness of the best rule from the 50 runs is 0.83 on training data and 0.61 on testing data. The rules that first identify low-permeability data have average fitness of 0.77 on training data and 0.6 on testing data. In both cases, there is a big gap between the fitness on training data and the fitness on testing data. There can be a couple of explanations. First, the two wells have very different geology.

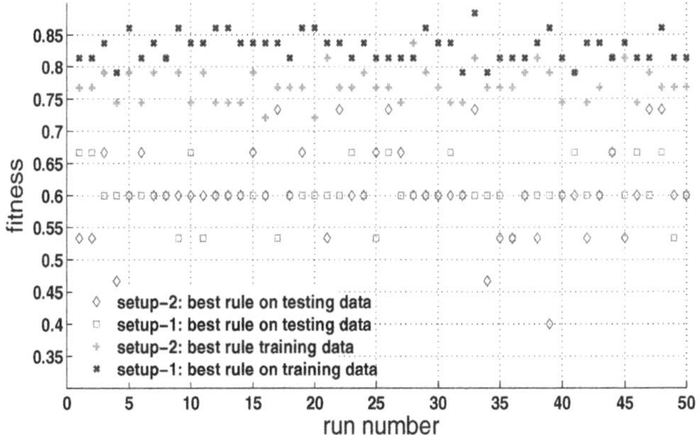

Figure 9-19.   Results of the two sets of runs.

The fuzzy rules trained based on log data from Well A therefore do not work as well on Well B. Another explanation is that Well B has a smaller number (15) of data points. Consequently, even a small number of mis-classification (1 or 2) will have strong impact on the classification accuracy. The accuracy measure of Well B, therefore, is not sufficient to reflect the fuzzy rules' performance.

To give a more detailed analysis of the performance of the fuzzy rules, we selected the rule that had the best fitness (0.76) on testing data and plotted its permeability interpretation on training data (Well A) and on testing data (Well B). The results are given in Figure 9-20 and Figure 9-21.

As shown, the fuzzy rule gives permeability interpretations which are very close to the transformed target permeability in both wells. In Well A, 8 out of the 43 segments were mis-classified; all of them have medium-permeability and the fuzzy rule mis-classified them as either low-permeability or high-permeability. The degree of 'mistake' is not too serious.

For Well B, the fuzzy rule mis-classified 4 out of the 15 segments. Among them, 1 segment can be fuzzily interpreted as either medium or high permeability according to the core permeability. The segmentation method transformed it as c (high permeability) while the fuzzy rule interpreted it as medium permeability. Once this segment is excluded, the number of mis-classification on Well B becomes 3 and the classification accuracy improves to 0.8, which is close to the accuracy on Well A (0.81). Based on this detailed analysis, the fuzzy rule gives a reasonably accurate permeability interpretation for both Well A and Well C. This is a very encouraging result.

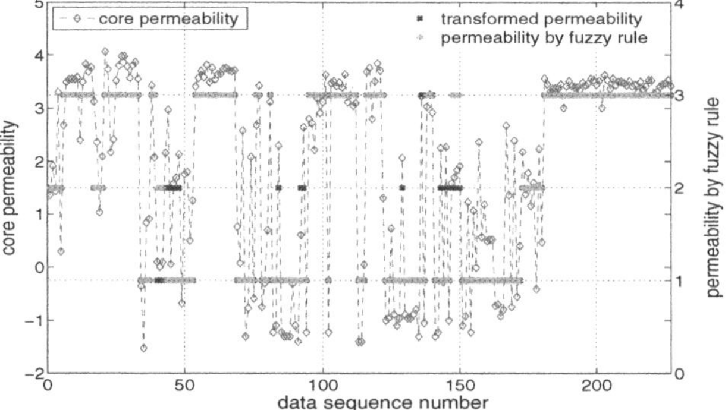

Figure 9-20.   Well A permeability.

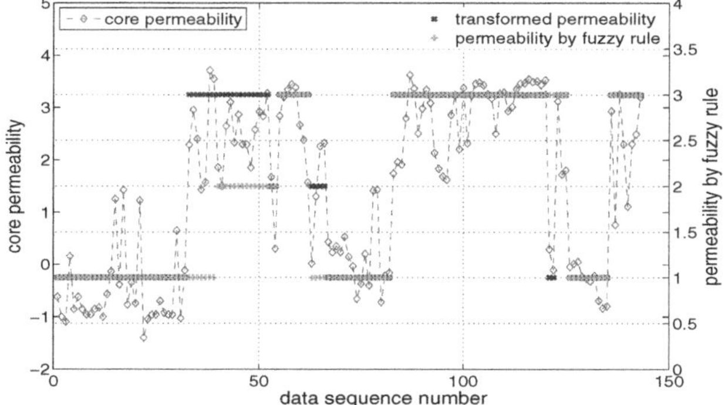

Figure 9-21.   Well B permeability.

7.   ANALYSIS AND DISCUSSION

To understand why rules that classify high-permeability segments first have produced better results, we calculated the average population fitness and the fitness of the best solution for all runs. The averages of the 50 runs for each set of experiments are plotted in Figure 9-22 and Figure 9-23.

When the first population is used to evolve rules that classify low-permeability segments and the second population is used to evolve rules that classify high-permeability data segments, Figure 9-22 shows that the co-evolution pressure is biased toward the second population. Average fitness of the first population is consistently lower than that of the second population. Using the worse of the two rules (the one from the first population) as *rule-1* to interpret permeability has impaired the overall interpretation accuracy.

This bias, however, does not appear in the other experiment where the rules that classify high-permeability were used as *rule-1* to interpret permeability. As shown in Figure 9-23, both populations co-evolve together with comparable average fitness. This is a healthy co-evolutionary dynamics which has produced combined if-then-else rule that give more accurate permeability interpretations than that by the other experiment.

In both sets of experimental runs, the two populations improved very quickly at the first 200 generations. After that, the improvement is not very visible. This pattern also appears in the fitness improvement of the best combined overall permeability interpreter, although to a lesser extend. One possible reason is

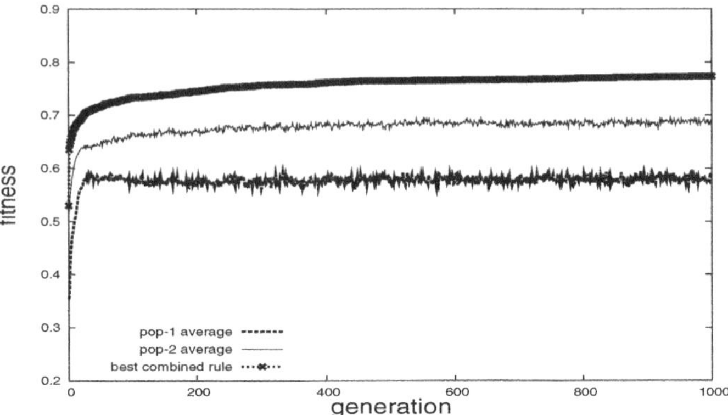

Figure 9-22.   Experimental results for runs where population 1 evolves rules to identify low-permeability data.

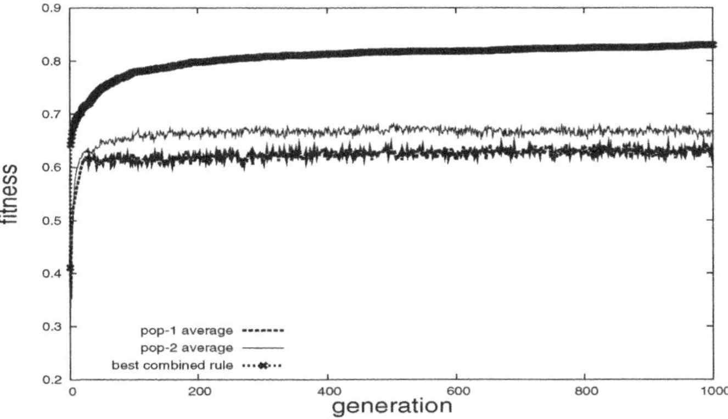

Figure 9-23.   Experimental results for runs where population 1 evolves rules to identify high-permeability data.

that the *best* solution used to combine with individuals in the other population for fitness evaluation is updated *every* generation. Such a greedy approach may have reduced the population diversity necessary for continuous evolution. In our future work, we plan to investigate using a less frequent updating scheme so that the two populations only occasionally communicate with each other. This asynchronous version of co-evolution model not only allows each population to have a slower and more stable evolution pace but also is suited for a parallel implementation in which each population is evolved on a separate processor. Such parallel implementation is important for the efficient processing of a large number of well logs simultaneously.

8.   CONCLUSIONS

Well log interpretation is a routine, but time-consuming task in oil companies. With the increasing global energy demand, it is a natural trend to seek computerized well log interpretation techniques to provide results more efficiently. In this work, we have devised a co-evolutionary fuzzy system to generate a well log interpreter to automatically process well log data and interpret reservoir permeability. The initial testing results show that the generated fuzzy rules give a sensible permeability interpretation. Although the result is preliminary, it provides initial evidence of the potential of the developed method. We plan to continue the work by extending the system in two areas:

- the capability to evolve fuzzy rule interpreters for other reservoir properties, such as lithology.

- a less frequent rule updating scheme between the two populations, hence the possibility of parallel implementation of the co-evolutionary fuzzy system.

Acknowledgments

I would like to thank Julian Squires for porting the PolyGP system from Haskell to Java. The data set is provided by Chevron Energy Technology Company.

References

Abonyi, J., Feil, B., Nemeth, S., and Arva, P. (2005). Modified gath-geva clustering for fuzzy segmentation of multivariate time-series. *Fuzzy Sets and Systems*, 149:39–56.

Apostolico, A., Bock, M. E., and Lonardi, S. (2002). Monotony of surprise and large-scale quest for unusual words. In *Proceedings of the 6th International Conference on Research in Computational Molecular Biology*, pages 22–31.

Bentley, Peter J. (2000). "Evolutionary, my dear watson" investigating committee-based evolution of fuzzy rules for the detection of suspicious insurance claims. In *Proceedings of the Genetic and Evolutionary Computation Conference (GECCO-2000)*, pages 702–709. Morgan Kaufmann.

Keogh, Eamonn, Chakrabarti, Kaushik, Mehrotra, Sharad, and Pazzani, Michael (2001). Locally adaptive dimensionality reduction for indexing large time series databases. In *Proceedings of ACM SIGMOD Conference on Management of Data*, pages 151–162.

Larsen, R. J. and Marx, M. L. (1986). *An Introduction to Mathematical Statistics and Its Applications, 2nd Edition*. Prentice Hall, Englewood.

Lin, J., Keogh, E., Lonardi, S., and Chiu, B. (2003). A symbolic representation of time series, with implications for streaming algorithms. In *Proceedings of the 8th ACM SIGMOD Workshop on Research Issues in Data Mining and Knowledge Discovery*.

Pena-Reyes, Carlos Andres and Sipper, Moshe (2001). Fuzzy coco: A cooperative coevolutionary approach to fuzzy modeling. *IEEE Transactions on Fuzzy Systems*, 9(5):727–737.

Potter, Mitchell A. and Jong, Kenneth A. De (1994). A cooperative coevolutionary approach to function optimization. In *Parallel Problem Solving from Nature – PPSN III*, pages 249–257, Berlin. Springer.

Potter, Mitchell A. and Jong, Kenneth A. De (2000). Cooperative coevolution: An architecture for evolving coadapted subcomponents. *Evolutionary Computation*, 8(1):1–29.

Yu, Tina (2001). Hierachical processing for evolving recursive and modular programs using higher order functions and lambda abstractions. *Genetic Programming and Evolvable Machines*, 2(4):345–380.

Yu, Tina and Bentley, Peter (1998). Methods to evolve legal phenotypes. In *Parallel Problem Solving from Nature – PPSN V*, pages 280–291, Berlin. Springer.

Yu, Tina and Wilkinson, Dave (2007). A fuzzy symbolic representation for intelligent reservoir well log interpretation. In "Hybrid Intelligent Systems using Soft Computing" of the Series on Computational Intelligence, Springer Verlag Edited by, O. Castillo, P. Melin, W. Pedrycz, and J. Kacprzyk.

Yu, Tina, Wilkinson, Dave, and Xie, Deyi (2003). A hybrid GP-fuzzy approach for reservoir characterization. In Riolo, Rick L. and Worzel, Bill, editors, *Genetic Programming Theory and Practise*, chapter 17, pages 271–290. Kluwer.

Chapter 10

RESOURCE SCHEDULING WITH PERMUTATION BASED REPRESENTATIONS: THREE APPLICATIONS

Darrell Whitley[1], Andrew Sutton[1], Adele Howe[1] and Laura Barbulescu[2]

[1] *Computer Science Department, Colorado State University, Fort Collins, CO 80523 USA*
[2] *Robotics Institute Carnegie Mellon University Pittsburgh, PA 15213 USA*

Abstract   Resource based scheduling using permutation based representations is reviewed. Permutation based representations are used in conjunction with genetic algorithms and local search algorithms for solving three very different scheduling problems. First, the Coors warehouse scheduling problem involves finding a permutation of customer orders that minimizes the average time that customers' orders spend at the loading docks while at the same time minimizing the running average inventory. Second, scheduling the Air Force Satellite Control Network (AFSCN) involves scheduling customer requests for contact time with a satellite via a ground station, where slot times on a ground station is the limited resource. The third application is scheduling the tracking of objects in space using ground based radar systems. Both satellites and debris in space must be tracked on regular basis to maintain knowledge about the location and orbit of the object. The ground based radar system is the limited resource, but unlike AFSCN scheduling, this application involves significant uncertainty.

Keywords: resource scheduling, genetic algorithms, local search, permutations, representation

1.   INTRODUCTION

The goal of resource scheduling is to allocate limited resources to requests during some period of time. A schedule may attempt to maximize the total number of requests that are filled, or the summed value of requests filled, or to optimize some other metric of resource utilization. Each request needs to be assigned a time window on some appropriate resource with suitable capabilities and capacity. Given the often large numbers of possible combinations of time

D. Whitley et al.: *Resource Scheduling with Permutation Based Representations*, Studies in Computational Intelligence (SCI) **88**, 219–243 (2008)
www.springerlink.com

slots and resources, a simple strategy is a greedy scheduler: allocate resources on a first-come-first-served basis by placing each request taken in some order on its best available resource at its best available time.

The problem with the simple greedy strategy is that requests are not independent – when one request is assigned a slot on a resource, that slot is not longer available. Thus, placing a single request at its optimal position may preclude the optimal placement of multiple other requests. A significant improvement then is to explore the space of possible permutations of requests where the permutation defines a priority ordering of the requests to be placed in the schedule.

Representing a schedule as a permutation offers several advantages over simply using a greedy scheduler alone or searching the schedule space directly. Permutations support a strong separation between the problem representation and the actual details of the particular scheduling application. This allows the use of relatively generic search operators that act on permutations and which are independent of the application. A more direct representation of the scheduling problem would require search operators that are customized to the application. When using a permutation based representation, this customization is hidden inside a greedy schedule builder.

Permutation representations do incur costs. First, the permutation is an indirect representation of the problem. Thus, a separate schedule builder must be constructed to map the permutation into an actual schedule for evaluation. How well the schedule builder exploits critical features of the problem may play a key role in how well the overall scheduling system works. Second, the permutation representation often introduces redundancies in the search space. Two different permutations may map to the same schedule. Assume request B is before request A in one permutation, but request A is before B in another permutation; otherwise the permutations are similar. If A and B do not compete for resources and otherwise do not interaction, the two permutations may map to exactly the same schedule. This would also mean that the two permutations also have the same evaluation. This redundancy can also contribute to the existence of plateaus, or connected regions with flat evaluation.

Permutation based representations were popularized because they support the application of genetic algorithms to resource scheduling problems. Whitley et al. (Whitley et al., 1989) first used a strict permutation based representation in conjunction with genetic algorithms for real world applications. However, Davis (Davis, 1985b) had previously used "an intermediary, encoded representation of schedules that is amenable to crossover operations, while employing a decoder that always yields legal solutions to the problem." This is also a strategy later adopted by Syswerda (Syswerda, 1991; Syswerda and Palmucci, 1991), which he enhanced by refining the set of available recombination operators.

The purpose of this chapter is to review how permutation representations have been used on three applications. The first example involves warehouse scheduling. The other two examples are from the related domains of satellite communication scheduling and radar tracking of objects (both satellites and debris) in space. In these last two examples, we examine the question of how well "heuristic search" works compared to optimal solutions obtained using exact methods. The problem with exact methods is that they can be costly. To make them work, one must sometimes decompose the problem. However, as will be shown, "heuristic" methods can yield superior results compared to "optimal exact methods."

For the applications, we will examine several factors related to the permutation representation. We will show how well the permutation representation performs in the applications. We will indicate how the domain information has been separated from the representation and embedded in the schedule builder. In some cases, we will show evidence of effects of the representation on the search space.

1.1 The Genitor Genetic Algorithm

The experiments discussed in this paper use the Genitor (Whitley, 1989) "steady-state" genetic algorithm (Davis, 1991). In the Genitor algorithm, two parents mate and produce a single child. The child then replaces the worst, or least fit, member of the population. Using a population of size P, the best P-1 individuals found during the search are retained in the population.

In addition, Genitor allocates reproduction opportunities based on the rank of the individuals in the population. A linear bias is used such that individuals that are above the median fitness have a rank-fitness greater than one and those below the median fitness have a rank-fitness of less than one. Assuming the population is sorted from best to worst, we will say that an individual has *rank* i if it is the ith individual in the sorted population. The selective pressure at i will be denoted $S(i)$ and corresponds to the number of times that the individual at rank i will be sampled in expectation over P samples under sampling with replacement. We will represent the overall selective pressure by $S = S(1)$, the selective pressure for the best individual in the population. For example, standard tournament selection has a linear selective pressure of $S = 2.0$ which implies the best individual in a population is sampled 2 times in expectation over P samples, while the median individual in the population is sampled 1 time, and the worst individual is sampled 0 times. Linear selective pressure can be implemented by a biased random number generator (see (Whitley, 1989)) or by using stochastic tournament selection. Stochastic tournament selection is implemented by comparing two individuals, but selecting the best with a probability greater than 0.5 but less than 1.0; if p_s is the probability of keeping the best individual, the selective pressure is then $2p_s$ (Goldberg, 1991).

1.1.1 Genetic Algorithm and Permutation Codings.

1.1.1   Genetic Algorithm and Permutation Codings.   Typically, simple genetic algorithms encode solutions using bit-strings, which enable the use of "standard" crossover operators such as one-point and two-point crossover (Goldberg, 1989). Some genetic algorithms also use real-valued representations (Davis, 1991).

When solutions for scheduling problems are encoded as permutations, a special crossover operator is required to ensure that the recombination of two parent permutations results in a child that (1) inherits good characteristics of both parents and (2) is still a permutation of the N task requests. Numerous crossover operators have been proposed for permutations representing scheduling problems.

Syswerda's (Syswerda, 1991) order crossover and position crossover differ from other permutation crossover operators such as Goldberg's PMX operator (Goldberg, 1985) or Davis' order crossover (Davis, 1985a) in that no contiguous block is directly passed to the offspring. Instead, several elements are randomly selected by absolute position. These operators are largely used for scheduling applications (e.g., (Syswerda, 1991; Watson et al., 1999; Syswerda and Palmucci, 1991) for Syswerda's operator) and are distinct from the permutation recombination operators that have been developed for the Traveling Salesman Problem (Nagata and Kobayashi, 1997; Whitley et al., 1989). Operators that work well for scheduling applications do not work well for the Traveling Salesman Problem, and operators that work well for the Traveling Salesman Problem do not work well for scheduling. Operators such as PMX and Cycle crossover represent early attempts to construct a general purpose permutation crossover operator; these have not been found to be well suited to scheduling applications.

Syswerda's order crossover operator can be seen as a generalization of Davis' order crossover (Davis, 1991) that also borrows from the concept of uniform crossover for bit strings. Syswerda's order crossover operator starts by selecting K uniform-random positions in Parent 2. The corresponding elements from Parent 2 are then located in Parent 1 and reordered so that they appear in the same relative order as they appear in Parent 2. Elements in Parent 1 that do not correspond to selected elements in Parent 2 are passed directly to the offspring.

```
               Parent 1:  "A B C D E F G"
               Parent 2:  "C F E B A D G"
      Selected Elements:     *   * *
```

For example, the selected elements in Parent 2 are F B and A in that order. A remapping operator reorders the relevant elements in Parent 1 in the same order found in Parent 2.

```
        "A B _ _ _ F _"  remaps to  "F B _ _ _ A _"
```

The other elements in Parent 1 are untouched, thus yielding

```
"F B C D E A G"
```

Syswerda also defined a "position crossover". Whitley and Nam (Whitley and Yoo, 1995) prove that Syswerda's order crossover and position crossover are identical in expectation when order crossover selects K positions and position crossover selects L-K positions over permutations of length L. In effect, order crossover inherits by order first, then fills the remaining slots by position. Position crossover inherits by position first, then fills the remaining slots by their relative order.

Syswerda's contribution (Syswerda, 1991) was to emphasize that permutation-based recombination operators can preserve either the position, relative order or adjacency of the elements in the parents when extracting information from the parents to construct a child. But operators cannot do all three things well. In various applications, we have found Syswerda's order crossover operator to be robust across a wide range of resource scheduling applications. This makes intuitive sense, given that the relative order in which requests are filled affects the availability of resources for later requests.

2. THE COORS WAREHOUSE SCHEDULING PROBLEM

The Coors production facility (circa 1990) consists of 16 production lines, a number of loading docks, and a warehouse for product inventory. At the time this research was originally carried out (Starkweather et al., 1991), each production line could manufacture approximately 500 distinct products; use of different packaging constitutes a different product. The plant contained 39 truck and 19 rail-car docks for loading customer orders. Orders could be filled directly from the production lines or from inventory.

A solution is a priority ordering of customer orders, given the mix of products that make up that order. A weekly production line schedule already exists.

Orders are separated into truck and rail-car orders before scheduling begins. A customer order remains at a dock until it is completely filled, at which point the dock becomes empty and available for another order. Only orders using an equivalent transport compete for dock space; however, all orders compete for product from either the production line or inventory.

In the schedule builder, orders are taken from the permutation of customer orders; in effect, the permutation queues up the customer orders which then wait for a vacant loading dock. When a dock becomes free, an order is removed from the queue and assigned to the dock. Orders remain at a dock until they are completely filled with product either drawn from inventory or directly from one of the production lines. Product comes from the production lines organized into pallets. An order specifies a method of transport (a truck or rail-car) as well as a certain combination of product pallets. During a typical 24 hour period, approximately 150 to 200 orders are filled.

Simulation is used to evaluate the quality of a particular schedule in this domain. We used both a fast *internal* simulator and a much slower high-resolution *external* simulator. The internal simulator executes in less than 0.01 second and provides evaluation information used by the objective function. The external simulator requires several minutes to execute. Thus, the search is done using the internal simulator.

The external simulator models warehouse events in much greater detail and was used to validate and confirm the quality of the schedules found by the search algorithms.

While the differences between the internal simulator and external simulator are largely a matter of detail, the differences can be important. For example, when product is moved from inventory to a loading dock in the internal simulator, the amount of time needed to move the product is a constant based on the average time required to move product to a particular dock. However, in the external simulator, individual fork lifts are modeled. The external simulator determines when the fork lift becomes available, where it is located and the route it needs to take to move the product. Over time, small differences in the two simulations can accumulate. However, to the degree that good schedules found using the internal simulator tend to be good schedules under the external simulator, search using the internal simulator can be effective. (We have also looked at other ways to combine and exploit both simulators during search (Watson et al., 1999).)

Figure 10-1 illustrates how a permutation is mapped to a schedule. Customer orders are assigned a dock based on the order in which they appear in the permutation; the permutation in effect acts as a customer priority queue. In the right-hand side of the illustration note that initially customer orders **A** to **I** get first access to the docks (in a left to right order). **C** finishes first, and the next order, **J**, replaces **C** at the dock. Order **A** finishes next and is replaced by **K**. **G** finishes next and is replaced by **L**.

If two customer orders need the same product, the customer order that has been at dock the longest gets the product first. Product is drawn from inventory or production on-demand, but with a bias toward product from the production line if it is available within a time horizon that does not impact total time at dock.

All results in this paper are based on an actual manufacturing plant configuration provided to us by Coors. The production line schedule and initial inventory were provided, as well as 525 actual customer orders filled over a three day period.

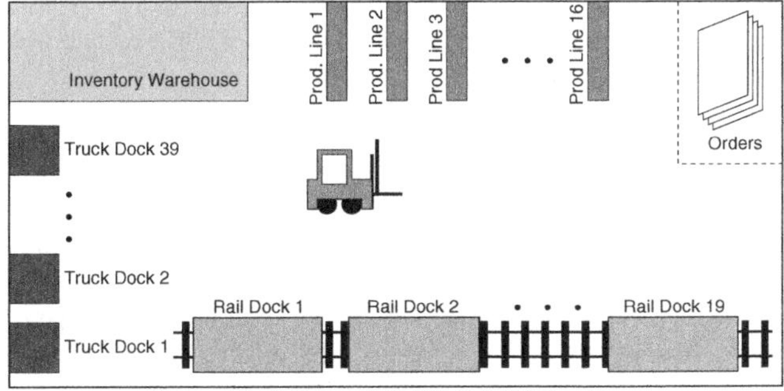

Customer Priority Queue: A, B, C, D, E, F, G, H, I, ..., Z

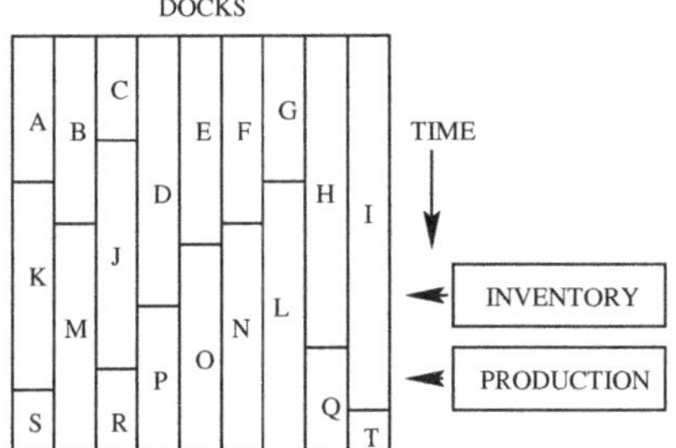

Figure 10-1. The warehouse model includes production lines, inventory, truck docks and rail docks. The columns in the schedule represent different docks. Customer orders are assigned to a dock left to right and product is drawn from inventory and the production lines.

2.1   The Coors Evaluation Function

For the Coors warehouse scheduling problem, we are interested in producing schedules that simultaneously achieve two goals. One of these is to minimize the mean time that customer orders remain at dock. Let N be the number of customer orders. Let M_i be the time that the truck or rail car holding customer order i spends at dock. Mean time at dock, $\mathcal{M}$, is then given by

$$\mathcal{M} = \frac{1}{N} \sum_{i=0}^{N} M_i.$$

The other goal is to minimize the running average inventory. Let F be the makespan of the schedule. Let $\mathcal{J}_t$ be inventory at time t. The running average inventory, I, is given by

$$I = \frac{1}{F} \sum_{t=0}^{F} \mathcal{J}_t.$$

Attempting to minimize either the mean time at dock or average inventory metrics independently can have a detrimental effect on the other metric. In this case, this multi-objective problem was transformed into a single-objective problem using a linear combination of the individual objectives proposed by Bresina (Bresina et al., 1995):

$$obj = \frac{(\mathcal{M} - \mu_{\mathcal{M}})}{\sigma_{\mathcal{M}}} + \frac{(I - \mu_I)}{\sigma_I} \tag{1}$$

where I represents running average inventory, $\mathcal{M}$ represents order mean time at dock, while μ and σ represent the respective means and standard deviations over a set of solutions.

2.2   Comparing Algorithm Performance

We have compared various algorithms to the results produced by the genetic algorithm for this problem (Watson et al., 1999). Here, we only report results for a stochastic hill-climber. The best results were given by an "exchange operator" (also referred to as a "swap operator"). This operator selects two random customers and then swaps their position in the permutation. We also report results using random sampling in conjunction with the greedy schedule builder. All of the algorithms reported here used 100,000 function evaluations. The Genitor algorithm used a population size of 500 and a selective pressure of 1.1.

For our test data, we have an actual customer order sequence developed and used by Coors personnel to fill customer orders. This solution produced an average inventory of 549817.25 product units and an order mean time at dock of 437.55 minutes. We consider a schedule *competitive* if these measures are less than or equal to those of the Coors solution. The first column of Table 10-1 illustrates that (mean) solutions obtained by random sampling are *not* competitive with the Coors solution. Random sampling in this situation does not imply random solutions; instead it involves the iterative application of the greedy schedule builder using 100,000 randomly generated permutations.

Results are presented in Table 10-1; we report mean performance and standard deviations over 30 runs. Statistical comparison of the competitive search algorithms indicates that both algorithms perform better than random sampling for both reducing inventory and time at dock, as verified by a series of two-tailed t-tests ($p < 0.0001$).

	Random Sampling		Genetic Algorithm		Exchange Hill Climber		Coors Solution
	Internal	External	Internal	External	Internal	External	Solution
Mean Time-At-Dock							
μ	456.88	447.11	392.49	397.48	400.14	439.43	437.55
σ	3.2149	8.0304	0.2746	7.3680	4.7493	4.0479	n.a.
Average Inventory							
μ	597123	621148	364080	389605	370458	433241	549817
σ	10041	28217	1715	9137	20674	20201	n.a.

Table 10-1. Performance Results on the Internal and External Simulator. The Coors Solution indicates a human generated solution used by Coors.

The final solution found using the internal simulator on each of 30 runs is also evaluated using the external simulator. Of course, the external simulator is expected to be a better predictor of actual performance. Comparing the actual events that occurred at Coors against the detailed external simulator, we find that Genitor was able to improve the mean time at dock by approximately 9 percent. It was the only scheduler that produced an improvement over what the human schedulers at Coors had done as measured by the external simulator. The big change however is in average inventory. Both Genitor and the local search methods show a marked reduction in average inventory. In the end, all of the schedules shipped exactly the same product; if two schedules finish shipping at the same time then the same amount of product must be left over. However, average inventory levels can be lower due to the combined effects of pulling strategic product from inventory early on and filling directly off of the production line. When time at dock is lower however, there is also an impact on inventory, since a longer overall schedule translates into a longer production period with the additional product going into inventory.

Overall, it is notable that the genetic algorithm produced results using the internal simulator that hold up rather well under the external simulator. The exchange hill climber did not fare as well under the external simulator. The genetic algorithm is the only scheduler with solutions for mean time at dock on the external simulator that improved on the human generated solution used by Coors.

One interesting question that could not be explored in this work was related to the long terms effects of less time at dock and lower average inventory. Are these gains sustainable over weeks and months as opposed to just 1 or 2 or 3 days? Could the warehouse be operated long-term with less overall inventory? These kinds of questions are not well addressed in the scheduling literature, and arise again in a later application in this paper.

Watson et al. (Watson et al., 1999) provide a more detailed discussion of the Coor's warehouse scheduling problem.

3.   SCHEDULING THE AIR FORCE SATELLITE CONTROL NETWORK

Scheduling the Air Force Satellite Control Network (AFSCN) involves coordinating communications via ground stations to more than 100 satellites. Space-ground communications are performed using 16 antennas located at nine ground stations around the globe. Customers submit requests to reserve an antenna at a ground station for a specified time period based on the visibility of the target satellites. We will separate the satellites into two types. The low altitude satellites have short (15 minutes) visibility windows that basically only allow one communication per pass over a ground station. High altitude satellites have longer windows of visibility that may allow for multiple tasks to be scheduled during a single pass over the ground station.

A problem instance consists of n task requests. A task request T_i, $1 \leq i \leq n$, specifies both a required processing duration T_i^{Dur} and a time window T_i^{Win} within which the duration must be allocated; we denote the lower and upper bounds of the time window by $T_i^{Win}(LB)$ and $T_i^{Win}(UB)$, respectively. Tasks cannot be preempted once processing is initiated. Each task request T_i specifies a resource (antenna) $R_i \in [1..m]$, where m is the total number of resources available. The tasks do not include priorities.

T_i may optionally specify $j \geq 0$ additional (R_i, T_i^{Win}) pairs, each identifying a particular alternative resource (antenna) and time window for the task. While requests are made for a specific antenna, often a different antenna at the same ground station may serve as an alternate because it has the same capabilities.

There are (at least) two approaches to defining evaluation functions for optimizing the utilization of the ground stations. One approach is to minimize the number of request conflicts for AFSCN scheduling; in other words we maximize the number of requests that can be scheduled without conflict. Requests that cannot be scheduled without conflict are bumped out of the schedule. This is historically the objective function that has been used for this problem.

However, this is not what happens when humans carry out AFSCN scheduling. Satellites are valuable resources, and the AFSCN operators work to fit in every request. So an alternative is to schedule every request, but minimize the amount of overlap in the schedule. Under this approach, all of the requests are scheduled, but some requests get less than the requested amount of time. In this approach, the amount by which requests must be trimmed to fit in every request (i.e., the overlap) is minimized.

To assess the relative merits of different heuristic search techniques on permutation representations of AFSCN, we compare performance of three algorithms: Genitor, local search and random sampling.

A genetic algorithm for minimizing conflicts searches permutations of contact requests. Genitor's schedule builder considers requests in the order that they appear in the permutations. Each task request is assigned to the first available resource (from its list of alternatives) and at the earliest possible starting time. If the request cannot be scheduled on any of the alternative resources, it is dropped from the schedule (i.e., bumped). The evaluation of a schedule is then defined as the total number of requests that are scheduled (for maximization) or inversely, the number of requests bumped from the schedule (for minimization).

Local search for minimizing conflicts uses the *shift* operator because we found it to work well compared to several relatively standard alternatives. From a current solution π, a neighborhood is defined by considering all $(N-1)^2$ pairs (x, y) of task request ID positions in π, subject to the restriction that $y \neq x - 1$. The neighbor π' corresponding to the position pair (x, y) is produced by *shifting* the job at position x into the position y, while leaving the relative order of other jobs unchanged.

Given the large neighborhood size, we use the shift operator in conjunction with next-descent hill-climbing: the neighbors of the current solution are examined in a random order, and the first neighbor with either a lower or equal number of bumped tasks is accepted. Search is initiated from a random permutation and terminates when a pre-specified number of solution evaluations is exceeded.

Random sampling for minimizing conflicts produces schedules by generating a random permutation of the task request IDs and evaluating the resulting permutation using the schedule builder. Randomly sampling a large number of permutations provides information about the distribution of solutions in the search space, as well as a baseline measure of problem difficulty for heuristic algorithms.

3.1 Results for Minimizing Conflicts

Parish (Parish, 1994) first applied the Genitor algorithm to AFSCN scheduling using Syswerda's order crossover with positive results. Parish used data from 1992 when about 300 requests were being scheduled each day. In the following experiments we used five days of actual data[1] for the dates: 3/7/2002, 3/20/2002, 3/26/2003, 4/2/2003 and 5/2/2003. The number of requests received during a typical day is approximately 450 each day. Our experiments show the increased demand from 300 to 450 or more request each day results in

---

[1] We thank William Szary and Brian Bayless at Schriever Air Force Base for providing us with data.

Day	Size	Genitor			Local Search			Random Sampling		
		Min	Mean	S.D.	Min	Mean	S.D.	Min	Mean	S.D.
03/07/02	483	42	43.7	0.98	68	75.3	4.9	73	78.16	1.53
03/20/02	457	29	29.3	0.46	49	56.06	3.83	52	57.6	1.67
03/26/03	426	17	17.63	0.49	34	38.63	3.74	38	41.1	1.15
04/02/03	431	28	28.03	0.18	41	48.5	3.59	48	50.8	0.96
05/02/03	419	12	12.03	0.18	15	17.56	1.3	25	27.63	0.96

Table 10-2. Performance of Genitor, local search and random sampling in terms of the best and mean number of bumped requests (with standard deviation as S.D.). All statistics are taken over 30 independent runs, with 8000 evaluations per run.

significantly more difficult problems because the available time on the ground stations is increasingly oversubscribed.

In Table 10-2, we present the results obtained for these problems. Statistics were obtained over 30 runs, with 8000 evaluations per run. We have run the algorithms for 100,000 evaluations, but the results change very little, and 8000 evaluations would allow human operators to run the schedulers interactively (the schedulers take less than 10 seconds to execute.)

3.2   A Hybrid Method with Optimal Low Altitude Scheduling

In our investigations we were able to prove that it is possible to schedule contacts with low altitude satellites optimally (Barbulescu et al., 2004). This is because the window of visibility for low altitude satellites typically allows only one contact per pass. This restricts the number of ways that a schedule can be constructed. Under these conditions we constructed a proof showing that a variant of activity selection scheduling (Corman et al., 1990) on multiple resources is optimal.

Thus, an alternative approach to this problem is to schedule the low altitude requests first using an exact method, and then schedule the remaining high altitude requests using a genetic algorithm or local search. We refer to this approach as the *the split heuristic*. In effect, we break the problem into two parts, and we know that one of the parts is solved optimally.

As shown in Table 10-3 the split heuristic fails to find the best known schedules for two problems, using Genitor with 8000 evaluations. The results remain the same even when dramatically more evaluations are used per run. By examining the data, we identified situations in which the low altitude requests were blocking insertion of longer high altitude requests.

Figure 10-2 illustrates a situation for which scheduling low-altitude requests first results in suboptimal solutions. Assume there are two ground stations and

Day	Best Known	Genitor-Split		
		Min	Mean	Stdev
03/07/02	42	42	42	0
03/20/02	29	30	30	0
03/26/03	17	18	18	0
04/02/03	28	28	28	0
05/02/03	12	12	12	0

Table 10-3. Results of running Genitor with the split heuristic over 30 experiments, with 8000 evaluations per experiment.

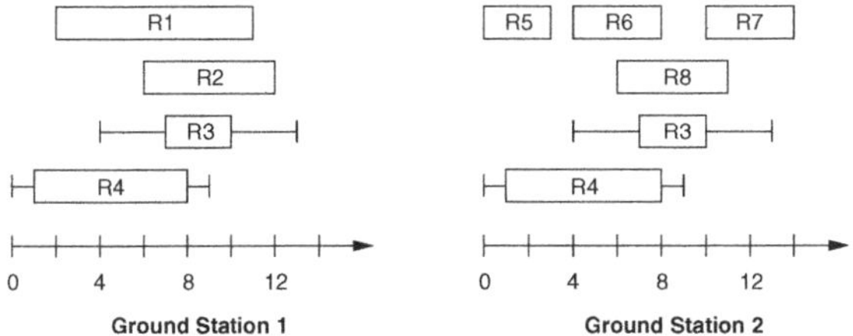

Figure 10-2. Example of a problem for which the split heuristic can not result in an optimal solution. Each ground station has two antennas; the only high-altitude requests are $R3$ and $R4$.

two resources (two antennas) at each ground station. Assume two high-altitude requests, $R3$ and $R4$, have durations three and seven, respectively. $R3$ can be scheduled between start time 4 and end time 13; $R4$ can be scheduled between 0 and 9. Both $R3$ and $R4$ can be scheduled at either of the two ground stations. The rest of the requests are low-altitude requests. $R1$ and $R2$ request the first ground station, while $R5$, $R6$, $R7$, and $R8$ request the second ground station.

If low-altitude requests are scheduled first, then $R1$ and $R2$ are scheduled on Ground Station 1 on the two resources, and the two high-altitude requests are bumped. Likewise, on Ground Station 2, the low-altitude requests are scheduled on the two resources, and the high-altitude requests are bumped. By scheduling low-altitude requests first, the two high-altitude requests are bumped. However, it is possible to schedule both of the high-altitude requests such that only one request ($R1$, $R2$ or $R8$) gets bumped. Therefore, an optimal solution is not possible when all of the low-altitude requests are scheduled before the high-altitude requests.

3.3   Minimizing Overlap

So far we have looked at the traditional objective of minimizing dropped requests (i.e., conflicts or bumps). Our second, more realistic objective is to minimize overlaps. The new objective function provides a richer evaluation function for the algorithm.

When minimizing the number of bumped tasks, if 500 jobs are being scheduled and at least 10 must be bumped, then the number bumped is always an integer between 10 and 500. If most of the time, the number of conflicts is between 10 and 100, then most of the time the evaluation function is an integer between 10 and 100. This means that the fitness landscape is made up of very large flat plateaus. Empirically we have verified that plateaus exist and have found that the plateaus are far too large to exhaustively enumerate. Thus the precise size of the plateaus is unknown. At an abstract level, this is very similar to the landscape induced by the classic MAXSAT optimization problems. Changing the location of two requests in the permutation sometimes (or even often) does not change the output of the evaluation function. Thus, the evaluation function is a coarse metric. And the key to finding better schedules involves finding "exits" off of the plateaus leading to a better solution.

When using overlaps as an evaluation, the evaluation function is not just related to the number of tasks that must be scheduled, but also to their durations. If the number of conflicts is between 1 and 100, but the overlaps range from 1 to 50 time units, then the evaluation function can range over 1 to 5000. The landscape is still dominated by plateaus; however, the evaluation functions provides more differentiation between alternative solutions. This translates into fewer and small plateaus in the search space.

Discussions with humans who schedule AFSCN by hand suggests that minimizing conflicts in not the correct goal. The goal is to fit in all of the requests, even if that means modifying some requests.

An example of how the sum of overlaps is computed is presented in Figure 10-3. Note that this is a solution for the example problem in Figure 10-2. R8 could either be scheduled on antenna A1 or antenna A2 at Ground Station 2. In order to minimize the overlaps, we schedule R8 on A1 (the sum of overlaps with R6 and R7 is smaller than the sum of overlaps with R3 and R4). While this is a trivial example, it illustrates the fact that instead of just reporting R8 as bumped, the new objective function results in a schedule which provides guidance about the fewest modifications needed to accommodate R8. Perhaps R6 and R7 and R8 can be trimmed slightly, or perhaps R6 and R7 can be shifted slightly outside of their request windows; this may be better than getting bumped entirely.

We designed a schedule builder for Genitor to schedule all tasks and compute the sum of the overlaps. If a request cannot be scheduled without conflict on

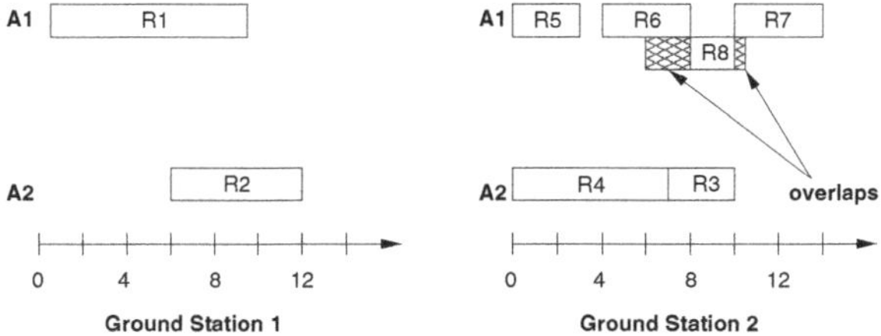

Figure 10-3. Optimizing the sum of overlaps.

any of the alternative resources, it overlaps; we assign such a request to the alternative resource on which the overlap with requests scheduled so far is minimized.

In Table 10-4, we present the results of running Genitor minimizing conflicts and Genitor minimizing overlaps for 30 runs, with 8000 evaluations per run. For "Genitor minimizing conflicts" we compute both the number of conflicts as well as the corresponding sum of overlaps (even though the schedule was not optimized to reduce overlaps) for the best schedule obtained in each run. Likewise, for "Genitor minimizing overlaps" we not only computed overlaps, we also computed the number of bumps needed to de-conflict the schedule.

The results show clearly that optimizing the number of conflicts results on average in a larger corresponding sum of overlaps than when the overlaps are optimized, and the increase can be quite significant. On the other hand, optimizing the sum of overlaps results in a number of bumps which is usually larger than when the conflicts are optimized; the increase is significant. These results also suggest that when minimizing the number of conflicts, longer tasks are bumped, thus resulting in a large sum of overlaps.

The significance of these results mainly relates to the correct choice of evaluation function. Using the wrong evaluation function can result in a scheduler that completely fails to solve the scheduling problem in a useful and appropriate manner. When conflicts are minimized, there is a strong bias toward removing large, difficult to schedule tasks. But if one must ultimately somehow put these back into the schedule it may require taking the whole schedule apart and starting over. Minimizing overlaps results in schedules that deal with all of the tasks that must be scheduled.

Barbulescu et al. (Barbulescu et al., 2004) provide a more detailed discussion of this problem. A more recent paper (Barbulescu et al., 2006) studies

| | Genitor minimizing conflicts | | | | | |
| | Conflicts | | | Overlaps | | |
Day	Min	Mean	S.D.	Min	Mean	S.D.
03/07/02	42	43.7	0.9	1441	1650.8	76.6
03/20/02	29	29.3	0.5	803	956.23	53.9
03/26/03	17	17.6	0.5	790	849.9	35.9
04/02/03	28	28.03	0.18	1069	1182.3	75.3
05/02/03	12	12.03	0.18	199	226.97	20.33

| | Genitor minimizing overlaps | | | | | |
| | Conflicts | | | Overlaps | | |
Day	Min	Mean	S.D.	Min	Mean	S.D.
03/07/02	55	61.4	2.9	913	987.8	40.8
03/20/02	33	39.2	1.9	519	540.7	13.3
03/26/03	24	27.4	10.8	275	292.3	10.9
04/02/03	35	38.07	1.98	738	755.43	10.26
05/02/03	12	12.1	0.4	146	146.53	1.94

Table 10-4. The results obtained for Genitor minimizing conflicts and Genitor minimizing overlaps by running 30 experiments with 8000 evaluations per experiment. The resulting schedules were then analyzed for both the number of bumps needed to de-conflict the schedule and the number of overlaps leaving everything in the schedule. This was done by using the permutation form of the solution in conjunction with the two different schedule builders.

several alternative methods for scheduling the AFSCN problem and explores why particular methods work well.

4. SCHEDULING WHEN TO TRACK OBJECTS IN SPACE

The Space Surveillance Network (SSN) is a collection of optical and radar sensor sites situated around the globe. The mission of the SSN is to maintain a catalog of thousands of objects in orbit around the earth. This catalog serves to facilitate object avoidance or contact, to prevent potential collisions during space flight, and to predict when specific orbits will decay.

The space catalog comprises both operational artificial satellites and debris from past space missions; each object is represented by information about its orbital trajectory. Many tracking sites in the SSN are *phased array* radars: devices that contain a two dimensional array of radio antennas that utilize constructive and destructive interference to focus a radiation pattern in a particular direction. Several objects can be tracked at once by interleaving track pulses subject to energy constraints. The *duty cycle* of the radar is a limit on the amount of energy available for simultaneously interleaved tracking tasks at a given instant. This defines the maximum resource capacity of the system. Due to surveillance and calibration tasks, the *available* resource capacity may fluctuate over time.

The SpaceTrack problem is an instance of SSN mission scheduling for an AN-FPS/85 phased array radar at Eglin Air Force Base in Florida, USA. Each day, radar operators at Eglin receive a consolidated list of requests (orbital objects with tracking frequency and durations) for performing observations on objects, along with their tracking priority. A *high* priority request must be filled during the day, while lower priority tasks are of less importance, and can be bumped in favor of critical requests. For each tracking request we identify a *priority weight* denoted by the scalar $w(r)$. The objective is to find an allocation that maximizes the priority-weighted sum of successful tracking tasks (WSS) subject to resource and temporal constraints.

It is important to note that this application involves uncertainty. Not every "track" that is scheduled will actually be successful. The exact size and shape of some objects may not be known. Irregularly shaped objects can display different radar cross sections from different perspectives. Thus, there are two estimations that create uncertainty: how much energy is needed to track the object, and the exact location of the object since its size and shape impacts it orbit. It is not uncommon for less than half of the scheduled tracks to be successful. When a track fails, it must be rescheduled.

With past information about an object's orbital trajectory and average radar cross section, we can use Keplerian laws to compute its expected *range* and *position* at a given time i. These expected quantities allow us to compute the following:

1. **Visibility windows.** The object's position at a given time defines whether it is visible to the array. Tracking cannot begin before the object rises over the local horizon (its *earliest start time* $est(i)$)and must complete before the object leaves the radar's field of view (its latest finish time $lft(i)$).

2. **Energy requirement** $c_j(i)$. The amount of energy necessary to illuminate an object is a function of the object's size and shape and its range with respect to the array. Distant objects require more energy than objects in low earth orbit. The sum of the energy requirements for all objects scheduled at i must not exceed a maximum available resource constraint (duty cycle limitation for the device).

3. **Expected value** $v_j(i)$. The probability of successfully tracking an object depends on how it is situated in the array's frame of reference. We derive the probability of detection by calculating an object's signal-to-noise ratio (SNR) profile over time. The SNR profile of an object is a function of its range and angle off the boresight direction (the normal vector to the

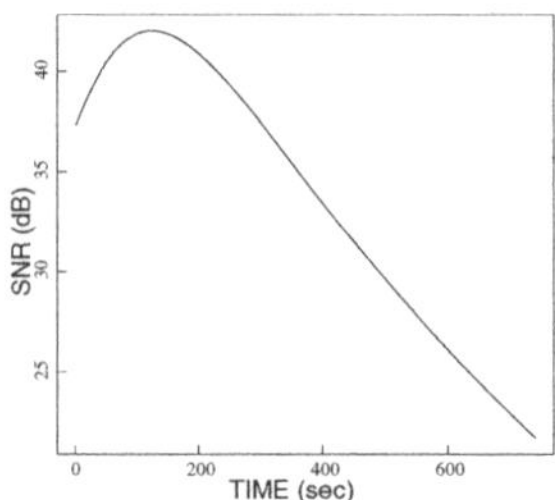

Figure 10-4. SNR profile of object with NORAD ID 27869 during a visibility window. The probability of detection (and expected weighted sum successful) is a monotonically increasing function of this quantity.

array's face plane). As the object moves through the array's field of view, this typically follows a smooth curve (see Figure 10-4). Multiplying this probability by the corresponding priority weight gives a tracking task's expected contribution to the weighted sum successful ($E[WSS]$).

Suppose system energy is infinitely available and the system has infinite capacity. In this case, the optimal solution can be calculated in polynomial time by assigning each task the tracking time that admits maximum SNR profile within a pass. However, when we constrain the available energy to realistic limits, the system becomes oversubscribed, and we can no longer feasibly schedule the entire set of tasks. Instead, we must construct a feasible subset of tasks to execute that maximizes the expected weighted sum successful.

On-peak scheduling.   One approach to constructing this subset would be to pick a collection of tasks and assign them their "peak" SNR profile time. In other words, the start times for each task are fixed at the value that gives peak SNR (we call this value $x^*(i)$) and thus the highest probability of detection. We will call this approach *on-peak* scheduling.

The on-peak approach can be characterized as an integer programming problem. Let n be the number of tasks that must be scheduled and τ denote the number of time units available.

$$\begin{aligned}\text{maximize} \quad & \mathbf{v}^T\mathbf{y} \\ \text{subject to} \quad & A\mathbf{y} \leq \mathbf{c} \\ & \mathbf{y} \in \{0,1\}^n\end{aligned} \qquad (2)$$

where $\mathbf{y}$ is a vector of n integer decision variables such that $\mathbf{y}_i = 1$ if the ith on-peak task is to be included, and $\mathbf{y}_i = 0$ otherwise. Similarly, $\mathbf{v}$ is a vector of

n real values such that $\mathbf{v}_i = v_{x^*(i)}(i)$, and $\mathbf{c}$ is a resource constraint vector of τ elements such that $\mathbf{c}_j = C_j$. Finally, A is the $\tau \times n$ constraint matrix defined as follows.

$$A_{j,i} = \begin{cases} c_j(i) & \text{if } x^*(i) \leq j < x^*(i) + d(i) - 1 \\ 0 & \text{otherwise} \end{cases} \quad (3)$$

This is exactly an instance of the NP-hard $\{0, 1\}$ multidimensional knapsack problem (Garey and Johnson, 1979). It contains the traditional (single-dimensional) $\{0, 1\}$ knapsack problem as a special case when the constraint matrix is a $1 \times n$ row vector.

Relaxed scheduling. The on-peak approach is guaranteed to give a solution in which each task is scheduled to execute during the time that maximizes the expected value. However, a solution with maximal *cumulative* expected value is not necessarily an *on-peak* solution. By shifting tasks off their peak (e.g., two tasks away from each other), other tasks may be squeezed in between. This means an individual task will be allocated a suboptimal expected value as a result of being placed away from its peak SNR profile; but more total tasks can be scheduled. We will call this approach *relaxed* scheduling.

4.1 Algorithms for Spacetrack Scheduling

Scheduling SpaceTrack is a difficult problem. Finding an optimal on-peak schedule is NP-hard, and finding an optimal relaxed schedule is APX-hard (Sutton, 2006). This makes heuristic search an attractive option for finding approximate solutions.

The current method being used to perform Spacetrack Scheduling is a simple greedy scheduler. We will look at three other ways to solve the problem that are 10 to 20 percent better than greedy.

An exact solution. Though SpaceTrack is NP-hard, we can still generate the optimal on-peak solution on small tractable instances using a branch-and-bound technique to solve the integer programming problem defined in Equation (2). This gives us a baseline with which to compare methods that use the relaxed method.

On-peak local search. Suppose $\mathfrak{S}$ is the set of all tasks under consideration. A solution to on-peak scheduling is a subset $S \subseteq \mathfrak{S}$ of tasks that are all scheduled feasibly on their peak times to maximize $E[WSS]$. The schedule builder places each task in order of the permutation as close to its peak as possible.

Starting from an initial ordering, the on-peak local search (OP-LS) produces neighboring candidate solutions using the *exchange operator* which randomly swaps the position of tasks in the permutation. Since we are maximizing, next-ascent local search is used: a reordering that results in equal or higher $E[WSS]$ (as produced by the schedule builder) is accepted as the new permutation. Since the normal exchange neighborhood is $O(N^2)$ we select candidates for the exchange operator randomly, which makes this approach a form of stochastic hill-climbing.

Evolving relaxed schedules.   We also employed the genetic algorithm Genitor (Whitley, 1989) to search the relaxed schedule space. In this case, an individual is represented by a permutation, but some other mechanism is needed to place a tracking task at a particular location in the schedule. We used a variant of greedy search to place a task at its on-peak location, or if that is not possible, at the best off-peak location possible. Fitness values are then computed by finding the $E[WSS]$ of the schedule produced by the schedule builder.

4.2   The SpaceTrack Results

Since SpaceTrack belongs to the class of NP-hard problems, we generate a set of 25 small tractable instances on which the optimal on-peak solution can be found with branch-and-bound. Each small instance is comprised of 50 tasks and 50 time units. The visibility windows were created by drawing from random distributions. The calculation of signal-to-noise ratio profiles requires expensive computation, and the distribution of the resulting values is difficult to control. Therefore, on the small set, the expected value curves are generated synthetically using a Gaussian curve.

To compare the algorithms on a large realistic problem, we create an instance using real data from NORAD's space database. This instance contains 5,000 tasks that must be assigned times from a set of 86400 time units (seconds during one day). The expected value curves and visibility times are created using actual position and velocity data computed using the standard SGP (Simplified General Perturbations) model (Hoots and Roehrich, 1980). These data are partitioned into 24 one-hour scheduling periods that are solved separately. The schedule found by each algorithm in a scheduling period is executed on a simulator that assigns success based on the probability function derived from SNR profile at the assigned tracking time.

The results from the set of small problems are reported in Table 10-5. The relaxed algorithm consistently finds higher mean values than the optimal on-peak solution found by branch-and-bound. We note that the On-peak Local Search algorithms (OP-LS) also found the same optimal solution as the branch-and-bound method in every case reported here.

The real-world problem is too large to be solved to optimality using branch-and-bound. Instead, we compare the relaxed algorithm with OP-LS as a

	On-peak			Relaxed	
instance	OPT	OP-LS		GENITOR	
		μ	σ	μ	σ
small 1	20.696170	20.696170	0.000000	21.474456	0.016000
small 2	21.453930	21.453930	0.000000	21.796775	0.004230
small 3	21.095790	21.095790	0.000000	21.452337	0.049200
small 4	18.106710	18.106710	0.000000	18.299666	0.069200
small 5	15.539030	15.539030	0.000000	16.024361	0.044300
small 6	15.622720	15.622720	0.000000	15.803709	0.018800
small 7	17.672190	17.672190	0.000000	17.892604	0.015300
small 8	16.115000	16.115000	0.000000	16.673159	0.084400
small 9	24.944260	24.944260	0.000000	25.269401	0.070200
small 10	18.481870	18.481870	0.000000	18.858413	0.034700
small 11	15.453560	15.453560	0.000000	16.217739	0.074900
small 12	16.274500	16.274500	0.000000	17.751006	0.178000

Table 10-5. A sample of results for all algorithms on the small set. The goal is to maximize the expected yield. Column two (OPT) indicates the optimal on-peak solution found by mixed-integer programming solver.

surrogate for the optimal on-peak solution. The results from each algorithm in each scheduling period on the real-world problem appear in Table 10-6. In this table, each expected weighted sum successful value found in a scheduling period is normalized by the sum of priority weights of all requests that need to be tracked in that period. Due to the stochastic nature of the simulation, the set of requests that require tracking in any given period may vary. We therefore report the normalized value in order to provide a fairer comparison between the algorithms.

	On-peak	Relaxed
period	OP-LS	GENITOR
1	0.4174375	0.4755850
2	0.2740852	0.3270757
3	0.1819694	0.2287325
4	0.4684505	0.5591947
5	0.3588590	0.4447416
6	0.2001081	0.2739436
7	0.3268069	0.4129994
8	0.4448630	0.5243267
9	0.4251086	0.4824034
10	0.3749262	0.4375215
11	0.4592485	0.5370493
12	0.4739717	0.5497464

Table 10-6. Expected weighted sum successful (normalized) found by each algorithm in each scheduling period using real-world data.

Again, the algorithms that employ the relaxed approach obtain a higher expected value than that obtained using the on-peak local search (OP-LS). Higher expected values correspond to a higher yield of successful tasks. For the real-world problem we can also model the potential gain in actual tasks successfully tracked. The number of tracking tasks determined to be successful by the simulator represents the "yield" of the schedules. We report these data for each algorithm found after simulation of the entire schedule day in Table 10-7.

	On-peak	Relaxed
priority	OP-LS	GENITOR
1	922 (52.15%)	1078 (60.97%)
2	1073 (48.57%)	1209 (54.73%)
3	1018 (40.32%)	1257 (49.78%)
4	817 (27.92%)	1032 (35.27%)
5	606 (15.53%)	699 (17.91%)
TOTAL	4436 (33.28%)	5275 (39.57%)

Table 10-7. Yield of successfully tracked passes by priority after the completion of all schedules for an entire day. The number in parentheses represents the percentage of successfully tracked passes out of those requested.

The results reported here are much better than the greedy methods currently in use. We are continuing to explore different approaches for solving the Space-Track Scheduling problem. Sutton et al. (Sutton et al., 2007) is the most recent paper; this paper explores the use of dynamic local search. It also explores trade-offs in terms of scheduling as many tracks as possible versus minimizing the mean time between tracks. Because unsuccessful tracks must be rescheduled, this problem also poses interesting issues related to the sustained long term performance of a scheduling system.

5.   CONCLUSIONS

Our results on three applications demonstrate the viability of using heuristic search (local search and genetic algorithms) in combination with permutation based representations. In each case, the heuristic search methods outperformed existing approaches to the applications.

Our results also show that heuristic methods can improve on "optimal" search methods if the optimal method must be applied to a smaller decomposed problem (and hence the solutions are not optimal for the full problem) or if a restricted version of the problem must be solved in order to guarantee optimality.

Exploring the permutation space rather than directly manipulating schedules appears to be an effective search strategy. Directly manipulating schedules means that operators must be much more customized to act on a particular type of schedules. When using a permutation representation, the specifics and constraints for a specific type of schedule are hidden in the schedule builder. When changing from one application to another only the schedule builder needs to change. And our experience suggests it is much easier and intuitive to *build* a feasible schedule using the permutation as a priority queue than to *modify* a schedule so as to define a neighborhood of transformations that systemically converts one schedule into another schedule. Thus, the use of a permutation representation allows a great deal of code reuse from one application to the next.

One concern with both the Coors Scheduler and the SpaceTrack scheduler is that looking at one day's worth of data (or a few days of data) does not adequately evaluate the true value of using an improved scheduling system. In the case of the Coors problem, the improved schedules may not be totally sustainable in the long run; part of the reduction in time at dock may be due to better scheduling that exploits surplus inventory. However, with more product going directly from production line to dock there must be some reduction in inventory (and some older product in inventory will have to be discarded). When this happens some of the apparent reduction in the time at dock may be lost. However, this only happens because the overall scheduling is better; the overall operation should still be more efficient with less product going to inventory.

On the SpaceTrack problem we expect just the opposite effect: we expect additional long term gains. This is because when an object is not tracked the importance of tracking that object increases. A lower priority object that is not tracked can become a higher priority object. The relaxed approach to scheduling SpaceTrack could result in fewer high priority objects over time.

Exploring the sustained long term impact of a scheduling method is an interesting area of research that is not particularly well addressed in the literature on scheduling, and which deserves more attention.

Acknowledgments

We wish to thank Patrick O'Kane and Richard Frank at ITT industries for their generosity and assistance regarding the Eglin SpaceTrack radar model. This research was partially supported by a grant from the Air Force Office of Scientific Research, Air Force Materiel Command, USAF, under grants number F49620-00-1-0144 and F49620-03-1-0233. The U.S. Government is authorized to reproduce and distribute reprints for Governmental purposes notwithstanding any copyright notation thereon. Funding was also provided by the Coors Brewing Company, Golden, Colorado.

References

L. Barbulescu, A.E. Howe, L.D. Whitley, and M. Roberts (2006) Understanding algorithm performance on an oversubscribed scheduling application. *Journal of Artificial Intelligence Research (JAIR)*.

L. Barbulescu, J.P. Watson, D. Whitley, and A. Howe (2004) Scheduling Space-Ground Communications for the Air Force Satellite Control Network. *Journal of Scheduling*.

John Bresina, Mark Drummond, and Keith Swanson (1995) Expected solution quality. In *Proceedings of the 14th International Joint Conference on Artificial Intelligence*.

T. Cormen, C. Leiserson, and R. Rivest (1990) *Introduction to Algorithms*. McGraw Hill, New York.

Lawrence Davis (1985) Applying Adaptive Algorithms to Epistatic Domains. In *Proc. IJCAI-85*.

Lawrence Davis (1985) Job Shop Scheduling with Genetic Algorithms. In John Grefenstette, editor, *Int'l. Conf. on GAs and Their Applications*, pages 136–140.

Lawrence Davis (1991) *Handbook of Genetic Algorithms*. Van Nostrand Reinhold, New York.

M. R. Garey and David S. Johnson (1979) *Computers and Intractability: A Guide to the Theory of NP-Completeness*. W. H. Freeman.

David Goldberg (1989) *Genetic Algorithms in Search, Optimization and Machine Learning*. Addison-Wesley, Reading, MA.

David Goldberg and Jr. Robert Lingle (1985) Alleles, Loci, and the Traveling Salesman Problem. In John Grefenstette, editor, *Int'l. Conf. on GAs and Their Applications*, pages 154–159.

D.E. Goldberg and M. Rudnick (1991) Genetic algorithms and the variance of fitness. *Complex Systems*, 5(3):265–278.

Felix R. Hoots and Ronald L. Roehrich (1980) Spacetrack report no. 3: Models for propagation of NORAD element sets. Technical report, Peterson Air Force Base.

Yuichi Nagata and Shigenobu Kobayashi (1997) Edge assembly crossover: A high-power genetic algorithm for the traveling salesman problem. In T. Bäck, editor, *Proc. of the 7th Int'l. Conf. on GAs*, pages 450–457. Morgan Kaufmann.

D.A. Parish (1994) A Genetic Algorithm Approach to Automating Satellite Range Scheduling. In *Masters Thesis*. Air Force Institute of Technology.

T. Starkweather, S. McDaniel, K. Mathias, D. Whitley, and C. Whitley (1991) A Comparison of Genetic Sequencing Operators. In L. Booker and R. Belew, editors, *Proc. of the 4th Int'l. Conf. on GAs*, pages 69–76. Morgan Kaufmann.

A. Sutton, A.E. Howe, and L.D. Whitley (2007) Using adaptive priority weighting to direct search in probabilistic scheduling. In *The International Conference on Automated Planning and Scheduling*.

A.M. Sutton (2006) *A two-phase dynamic local search algorithm for maximizing expected success in on-line orbit track scheduling*. MS thesis, Colorado State University, Fort Collins, CO.

Gilbert Syswerda (1991) Schedule Optimization Using Genetic Algorithms. In Lawrence Davis, editor, *Handbook of Genetic Algorithms*, chapter 21. Van Nostrand Reinhold, New York.

Gilbert Syswerda and Jeff Palmucci (1991) The Application of Genetic Algorithms to Resource Scheduling. In L. Booker and R. Belew, editors, *Proc. of the 4th Int'l. Conf. on GAs*. Morgan Kaufmann.

J.P. Watson, S. Rana, D. Whitley, and A. Howe (1999) The Impact of Approximate Evaluation on the Performance of Search Algorithms for Warehouse Scheduling. *Journal on Scheduling*, 2(2):79–98.

Darrell Whitley, Timothy Starkweather, and D'ann Fuquay (1989) Scheduling Problems and Traveling Salesmen: The Genetic Edge Recombination Operator. In J.D. Schaffer, editor, *Proc. of the 3rd Int'l. Conf. on GAs*. Morgan Kaufmann.

Darrell Whitley and Nam-Wook Yoo (1995) Modeling Permutation Encodings in Simple Genetic Algorithm. In D. Whitley and M. Vose, editors, *FOGA - 3*. Morgan Kaufmann.

L. Darrell Whitley (1989) The GENITOR Algorithm and Selective Pressure: Why Rank Based Allocation of Reproductive Trials is Best. In J.D. Schaffer, editor, *Proc. of the 3rd Int'l. Conf. on GAs*, pages 116–121. Morgan Kaufmann.

Chapter 11

EVOLUTIONARY COMPUTATION IN THE CHEMICAL INDUSTRY

Arthur Kordon
The Dow Chemical Company, Freeport, TX, U.S.A.

Abstract   Evolutionary computation has created significant value in the chemical industry by improving manufacturing processes and accelerating new product discovery. The key competitive advantages of evolutionary computation, based on industrial applications in the chemical industry are defined as: no a priori modeling assumptions, high quality empirical models, easy integration in existing industrial work processes, minimal training of the final user, and low total cost of development, deployment, and maintenance. An overview of the key technical, organizational, and political issues that need to be resolved for successful application of EC in industry is given in the chapter. Examples of successful application areas are: inferential sensors, empirical emulators of mechanistic models, accelerated new product development, complex process optimization, effective industrial design of experiments, and spectroscopy.

Keywords: evolutionary computation, competitive advantage, industrial applications, chemical industry, application issues of evolutionary computation.

1.   INTRODUCTION

The business potential of Evolutionary Computation (EC) in the area of engineering design was rapidly recognized in the early 1990s by companies like GE, Rolls Royce, and British Aerospace (Parmee, 2001). In a short period of several years, many industries, such as aerospace, power, chemical, etc., transferred their research interest for EC into various practical solutions. Contrary to its very nature, EC entered the industry in a revolutionary rather than an evolutionary way.

However, the pressure for more effective innovation strategies has significantly changed the current Research and Development environment in industry (Christensen et al., 2004). One of the key consequences is the expected reduced

A. Kordon: *Evolutionary Computation in the Chemical Industry*, Studies in Computational Intelligence (SCI) **88**, 245–262 (2008)
www.springerlink.com   © Springer-Verlag Berlin Heidelberg 2008

cost of technology exploration. As a result, priorities are given to those methods which can deliver practical solutions with minimal investigation and development efforts. This requirement not only pushes towards a better understanding of the unique technical capabilities of a specific approach, but also focuses attention on evaluating the total cost-of-ownership (potential internal research, software development and maintenance, training, and implementation efforts, etc.). The last task is not trivial, especially in an environment where several diverse approaches, such as first-principles modeling, statistical modeling, neural networks, fuzzy systems, support vector machines, etc, are competing with each other.

Another critical task in applying emerging technologies, such as EC, in industry is analyzing and resolving the key technical and non-technical issues that may slow down or even stop the implementation efforts. Of special importance are the non-technical issues, due to the unfamiliarity of the technology in the majority of the industry.

The objective of this chapter is to present the current state of the art of applied EC in the chemical industry, based on the experiences and implementation of this technology in The Dow Chemical Company. The structure of the chapter is as follows: First, the specific needs for EC in the chemical industry are identified in Section 2. The importance of clarifying the competitive advantages for successful marketing and industrial applications of EC is discussed in Section 3. The key competitive advantages of EC, based on the existing applications in the chemical industry, are briefly presented in Section 4. Some technical, organizational and political challenges in applying EC are discussed in Section 5. Several application areas for EC in the chemical industry with demonstrated competitive advantage are defined in Section 6. Finally, Section 7 summaries the chapter.

2.   WHY DOES THE CHEMICAL INDUSTRY NEED EVOLUTIONARY COMPUTATION?

The chemical industry is the backbone of contemporary economy with annual global revenue of two trillion dollars. The Dow Chemical Company is the largest US chemical company by sales ($46 billion in 2005). It is one of the most diversified global companies in the business with more than 3000 products in the market, mostly plastics, polyurethanes, and different specialty chemicals. There are two key directions in introducing new emerging technologies, such as EC, in the company. The first direction is continuous improvement of manufacturing and supply-chain efficiency, which becomes critical in the current economic environment of sky-rocketing energy prices. The second direction is faster invention of an attractive pipeline of new products, which is critical for the future competitiveness of the company.

The more specific issues that the introduced new technologies need to address can be defined as:

- high-dimensionality of plant data (thousands of variables and control loops)

- scarcity of data in new product development

- low speed of data analysis in High-Throughput Screening (which is at the basis of new product discoveries)

- increased requirements for model robustness toward process changes due to the cyclical nature of the industry

- multiple optimal solutions

- key process knowledge is owned by process operators and is poorly documented

- high uncertainty in new material market response

- supply chain is not as developed as process modeling.

There are tremendous opportunities for EC to satisfy these needs and to contribute to the process improvement and to discover new products. However, we need to take into account that as a result of the intensive modeling efforts in the last 20 years many manufacturing processes in the most profitable plants are already supplied by different types of models (steady-state, dynamic, model predictive control, etc.). This creates a culture of modeling fatigue and resistance to introduction of new solutions, especially based on unknown technologies. This makes the efforts of applying EC in the chemical industry especially challenging since demonstration of significant competitive advantages relative to the alternative modeling and hardware solutions is required.

3. IMPORTANCE OF DEFINING COMPETITIVE ADVANTAGES FOR INDUSTRIAL APPLICATIONS

Webster's Dictionary defines the term "advantage" as the superiority of position or condition, or a benefit resulting from some course of action. "Competitive" also relates to, characterized by, or based on competition (rivalry). A competitive advantage can result either from implementing a value-creating strategy not simultaneously being employed by current or prospective competitors or through superior execution of the same strategy as competitors (Bharadwaj et al., 1993). Based on this economic characterization, we define the competitive advantage of a research approach as technical superiority that cannot be reproduced by other technologies and can be translated with minimal efforts into a position of value creation in the marketplace. The definition includes three components. The first component requires clarifying the technical superiority (for example, better predictive accuracy) over alternative methods. The second component assumes minimal total cost-of-ownership and requires assessment of the potential implementation efforts of the competitive technologies. The third component is based on the hypothesis that the technical gain can improve the business performance and contribute to economic competitive advantage.

The technical advantages can be evaluated by benchmarking or comparative analysis (see examples for neural networks, genetic algorithms, and fuzzy systems in (Jain and Martin, 1999). However, the assessment of the total cost-of-ownership and the potential translation of the technical superiority into competitive products is still in its infancy. To our knowledge, there are no known public references on this topic. In this chapter, we define the competitive advantages of EC based on our experience in applying this technology in the chemical industry. Similar analysis for other industries is needed for more generic conclusions.

4.   COMPETITIVE ADVANTAGES OF EC IN CHEMICAL INDUSTRY

EC has been applied in different businesses and technical areas in The Dow Chemical Company since the early 90s (Kotanchek et al., 2002; Kordon et al., 2005). From our experience, one generic area where EC has demonstrated a clear competitive advantage is the development of simple empirical solutions in terms of models and rules. We have shown in several cases that the models, generated by EC are a low-cost alternative to both high fidelity models (Kordon et al., 2003b) and expensive hardware analyzers (Kordon et al., 2003a). The specific competitive advantages of EC related to the generic area of empirical modeling are defined as follows:

4.1   No A Priori Modeling Assumptions

EC models are developed with a minimal number of assumptions unlike, for example, first principles models that have many assumptions stemming from physical considerations or by statistical considerations, such as variable independence, multivariate normal distribution and independent errors with zero mean and constant variance. This assumption liberation establishes a technical superiority of generating models from the data with minimal effort from the experts[1]. The cost savings are in the experts' reduced time for defining and especially for validating the model assumptions. In case of mechanistic models

---

[1] However, all necessary data preparation procedures, such as data cleaning, dealing with missing data, outlier removal, etc. are still necessary.

for chemical processes, which may require defining and validating the assumption space of hundreds of parameters by several experts, the savings could be significant. On the other hand, in estimating the total cost-of-ownership, we have to take into account the additional time from the experts to select and interpret the generated assumption-free models.

4.2 High Quality Empirical Models

The key EC approach for empirical model building is symbolic regression, generated by Genetic Programming (GP). A well-known issue of the conventional GP algorithm, however, is the high complexity of the generated expressions at the end of the simulated evolution. In most of the cases the high fitness is based on very inefficient structure due to the generation of useless sub-trees, called introns (Banzhaf et al., 1998). The breakthrough method to resolve this issue is multi-objective simulated evolution where in addition to the performance as a first objective, the complexity of the generated symbolic regression expression is explicitly used as a second objective. In this case the optimal models fall on the curve of the nondominant solutions, called the Pareto front, i.e., no other solution is better than the solutions on the Pareto front in both complexity and performance (Smits and Kotanchek, 2004). Pareto front optimized GP allows the simulated evolution and model selection to be directed toward structures based on an optimal balance between accuracy and expression complexity. A current survey from several industrial applications in The Dow Chemical Company demonstrates that the selected final models are with very low level of complexity (Kordon et al., 2005). The derived symbolic regression models show improved robustness during process changes relative to conventional GP as well as neural network-based models.

4.3 Easy Integration in Existing Work Processes

In order to improve efficiency and reduce implementation cost, the procedures for model development, implementation, and maintenance in industry are standardized by work processes and methodologies. For example, many companies use the Six Sigma and Design for Six Sigma methodologies to operate their processes and introduce new products more efficiently (Breyfogel III, 2003). One of the positive effects of Six Sigma is the widespread use of statistical methods not only in empirical model building by the engineers but also in making statistically-based decisions by managers. Since the industry has already invested in developing and supporting the infrastructure of the existing work processes, the integration efforts of any new technology become a critical issue. From that perspective, EC in general and symbolic regression in particular, have a definite competitive advantage. We have shown that EC technology could be integrated within Six Sigma with minimal efforts as an extension of the existing statistical modeling capabilities. These additional nonlinear modeling capabilities are in a form of explicit mathematical expressions. Another advantage of this type of solutions is that there is no need for a specialized software environment for their run-time implementation (as is the case of mechanistic and neural network models). This feature allows for a relatively easy software integration of this specific EC technology into most of the existing model deployment software environments.

4.4 Minimal Training of the Final User

The symbolic regression nature of the final solutions, generated by GP, is universally acceptable by any user with mathematical background at a high school level. This is not the case, either with the first-principles models (where specific physical knowledge is required) or with the black-box models (where for example, some advanced knowledge on neural networks is a must). In addition, a very important factor in favor of symbolic regression is that process engineers prefer mathematical expressions and very often can find an appropriate physical interpretation. They usually don't hide their distaste toward black boxes.

4.5 Low Total Development, Deployment, and Maintenance Cost

The development cost includes expenses from the internal research efforts, internal software development efforts, and the research-type of marketing efforts to sell the technology to industrial users. From our experience, the internal research efforts to evaluate the capabilities of EC are comparable with other similar approaches, such as neural networks. Since we estimated and demonstrated the potential value impact of EC in the early phase of the internal research efforts, we were in a position to allocate resources to develop our own software and to improve the technology. This obviously added significantly to the development cost, although it was a very good investment.

EC has a clear advantage in marketing the technology to potential users. The scientific principles are easy to explain to almost any audience. We also find that process engineers are much more open to take the risk to implement symbolic regression models in the manufacturing plant rather than the alternatives because they can often relate the symbolic regression model to their chemical and physical understanding of the processes.

Most of the alternative approaches have a high cost of deployment, especially in real-time process monitoring and control systems. The license fee of the available on-line versions of the software is at least an order of magnitude more expensive than the off-line development option. As was discussed earlier, symbolic regression models do not require special run-time versions of the software and can be directly implemented in most existing process monitoring and control system, i.e. the deployment cost is minimal.

Another deployment problem where EC can reduce cost is in improving on the slow execution speed of some of the complex first-principle models. Very often, such models require at least 30-60 minutes of calculation time for selected operating conditions and this prevents them from real-time applications. One possible way to solve this problem and significantly speed up on-line execution time is by representing the complex mechanistic model with a set of empirical models, called emulators, generated by GP.

Often maintenance and support cost of applied models is neglected in the initial total cost estimates. It turns out, that this may take the lion's share of the total cost-of-ownership. For example, first-principles models require special-ized knowledge for their support. Model validation in continuously changing operating conditions becomes very time consuming and costly. Often the new operating conditions are outside of the assumption space and the validity of the model becomes questionable.

The changing operating conditions are even a bigger challenge for neural networks-based models and lead to frequent retraining and even completely new model redesign. As a result, both complex mechanistic models and neural net-works gradually become a maintenance nightmare. The growing maintenance cost may also bring into question the value of the model and lead to the decision of removing it from the process.

In contrast, the simple symbolic regression models require minimal main-tenance. From our experience the model redesign is very rare and most of the models perform with acceptable quality even when used for analysis as far as 20% outside their original model data range. There are symbolic regression models that have been in operation since 1997 (Kordon and Smits, 2001).

5.   APPLICATION ISSUES WITH EVOLUTIONARY COMPUTATION

Applying a new technology, such as EC, in industry requires not only re-solving many technical issues, but also handling problems of a non-technical nature systematically and patiently. A short overview of the key technical and non-technical issues is given below.

5.1   Technical Issues

- **Available computer infrastructure** - EC is one of the scientific ap-proaches that benefits directly from the fast growth of computational power. However, even with the help of Moore's Law, EC model devel-opment requires significant computational efforts. It is recommended to allocate a proper infrastructure, such as computer clusters, to accelerate this process. The growing capability of grid computing to handle computationally intensive tasks is another option to improve the EC performance, especially in a big global corporation with thousand of computers. However, development of parallel EC algorithms in user-friendly software is needed.

- **Professional EC software** - the current software options for EC implementation, either external or internally developed, are still used mostly for algorithm development and research purposes. One of the obstacles for potential mass scale applications of EC is the lack of professional user-friendly software from well-established vendors, which will take care also of continuous product development and support. Without such a product, the implementation effort is very high and it will be very difficult to leverage a consistent methodology for EC industrial applications.

- **Technical limitations of EC** - in spite of the fast theoretical development since the early 90s and increasing computational speed, EC still has several well-known limitations. First, generating solutions in a high-dimensional search space takes a significant amount of time. Second, model selection is not trivial and still more of an art than a science. Last, integrating heuristics and prior knowledge is not at the desirable level for practical applications. Generating complex dynamic systems by EC is still in its infancy.

5.2   Non-technical Issues

The ultimate success of an industrial application requires support from and interaction with multiple people (operators, engineers, managers, researchers, etc.). Because it may interfere with the interests of some of these participants, human factors, even politics, can play a significant role in the success of EC (Kordon et al., 2001). On the one hand, there are many proponents of the benefits of EC who are ready to cooperate and take the risk to implement this new technology. Their firm commitment and enthusiasm are a decisive factor to drive the implementation to the final success. On the other hand, however, there are also a lot of skeptics who do not believe in the real value of EC and look at the implementation efforts as a research toy exercise. In order to address these issues a well defined systematic approach in organizing the development and application efforts is recommended. In addition to the key technical issues, discussed in the previous section, the following organizational and political issues need to be resolved.

Organizational issues include the following:

- Critical mass of developers: It is very important at the early phase of industrial applications of EC to consolidate the development efforts. The probability for success based only on individual attempts is very low. The best-case scenario is to create a virtual group that includes not only specialists, directly involved in EC development and implementation, but also specialists with complementary areas of expertise like machine learning, expert systems, and statistics.

- EC marketing to business and research communities: Since EC is virtually unknown not only to business-related users but to most of the other research communities, it is necessary to promote the approach by significant marketing efforts. Usually this research-type of marketing includes series of promotional meetings based on two different presentations. The one directed toward the research communities focusing on the technology kitchen, i.e., gives enough technical details to describe the EC technologies, demonstrates the differences from other known methods, and clearly illustrates their competitive advantages. The other one targeting business-related audience focuses on the technology dishes, i.e., it demonstrates with specific industrial examples the types of applications that are appropriate for EC, describes the work process to develop, deploy, and support an EC application, and illustrates the potential financial benefits.

- Link EC to proper corporate initiatives: The best case strategy we recommend is to integrate the development and implementation efforts within the infrastructure of a proper corporate initiative. A typical example is the Six Sigma initiative, which is practically a global industrial standard. In this case the organizational efforts will be minimized since the companies have already invested in the Six Sigma infrastructure and the implementation process is standard and well known.

Political issues include the following:

- Management support: Consistent management support for at least several years is critical for introducing any emerging technology, including EC. The best way to win this support is to define the expected research efforts and to assess the potential benefits from specific application areas. Of decisive importance, however, is the demonstration of any value creation by resolving practical problems as soon as possible.

- Skepticism and resistance toward EC technologies: There are two key sources with this attitude. The first source is the potential user in the businesses who is economically pushed more and more toward inexpensive, reliable, and easy-to-maintain solutions. In principle, EC applications require some training and significant cultural change. Many users are reluctant to take the risk even if they see a clear technical advantage.

A persistent dialog, supported with economic arguments and examples of successful industrial applications is needed. Sometimes sharing the risk by absorbing the development cost by the research organization is a good strategy, especially in applications that could be easily leveraged. The second source of skepticism and resistance is in the research community itself. For example, the majority of model developers prefers the first principles approaches and very often treats the data-driven methods as black magic which cannot replace solid science. In addition, many statisticians express serious doubts about some of the technical claims in EC technologies. The only winning strategy to change this attitude is by more intensive dialog and finding areas of common interest. An example of a fruitful collaboration between fundamental model developers and EC developers is given in the next section and described in (Kordon et al., 2002). Recently, a joint effort between statisticians and EC developers demonstrated significant improvement in using genetic programming in industrial statistical model building (Castillo et al., 2004).

- Lack of initial credibility: As a relatively new approach to industry, EC does not have a credible application history for convincing a potential industrial user. Almost any EC application requires a high-risk culture and significant communication efforts. The successful application areas, discussed in this paper, are a good start to gain credibility and increase the EC potential customer base in the chemical industry.

6.   KEY EVOLUTIONARY COMPUTATION APPLICATION AREAS IN THE CHEMICAL INDUSTRY

Based on our recent experience from several applications on different real industrial processes in the Dow Chemical Company, we would recommend the following industrial problems as appropriate for implementing EC in the chemical industry:

6.1   Fast Development of Nonlinear Empirical Models

Symbolic-regression types of models are very well-fit for industrial applications and are often at the economic optimum of development and maintenance cost. One area with tremendous potential is inferential or soft sensors, i.e. empirical models that infer difficult-to-measure process parameters, such as NOx emissions, melt index, interface level, etc., from easy-to-measure process variables such as temperatures, pressures, flows, etc. (Kordon et al., 2003b). The current solutions in the market, based on neural networks, require frequent retraining and specialized run-time software.

An example of an inferential sensor for a quality parameter in a distillation column prediction based on an ensemble of four different models is given in (Jordaan et al., 2004). The models were developed from an initial large manufacturing data set of 23 potential input variables and 6900 data points. The size of the data set was reduced by variable selection to 7 significant inputs and the models were generated by five independent GP runs. As a result of the model selection, a list of 12 models on the Pareto front was proposed for further evaluation to process engineers. All selected models have high performance (R^2 of 0.97-0.98) and low complexity. After evaluating their extrapolation capabilities with What-If scenarios, the diversity of model inputs, and by physical considerations, an ensemble of four models was selected for on-line implementation. Two of the models are shown below:

$$GPModel_1 = A + B\left(\frac{Tray64T^4 * Vapor^3}{RflxFlow^2}\right) \tag{1}$$

$$GPModel_2 = C + D\left(\frac{Feed^3 \sqrt{Tray46T - Tray56T*}}{Vapor^2 * RflxFow^4}\right) \tag{2}$$

where A, B, C, and D are fitted parameters, and all model inputs in the equations are continuous process measurements.

The models are simple and interpretable by process engineers. The different inputs used in both models increases the robustness of the estimation scheme in case of possible input sensor failure. The inferential sensor has been in operation since May 2004.

6.2 Emulation of Complex First-Principles Models

Symbolic regression models can be a substitute for parts of fundamental models for on-line monitoring and optimization. The execution speed of the majority of the complex first-principles models is too slow for real time operation. One effective solution is to emulate a portion of the fundamental model, by a symbolic regression model, called an emulator, built only with selected variables, related to process optimization. The data for the emulator are generated by design of experiments from the first-principles model. One interesting benefit of emulators is that they can be used as fundamental model validation indicators as well. Complex model validation during continuous process changes requires tremendous efforts in data collection and numerous model parameter fittings. It is much easier to validate the simple emulators and to infer the state of the complex model on the basis of the high correlation between them. An example of such an application for optimal handling of by-products is given in (Kordon et al., 2003a). The mechanistic model is very complex and includes over 1500 chemical reactions with more than 200 species. Ten input variables and 12 output variables that need to be predicted and used in process optimization were selected from the experts. A data set, based on a four level design of experiments, was generated and used for model development and validation. For 7 of the outputs, a linear emulator gave acceptable performance. For the remaining 5 outputs, a nonlinear model was derived by GP. An example of a nonlinear emulator, selected by the experts is given below:

$$Y_5 = \frac{6x_3 + x_4 + x_5 + 2x_6 + x_2x_9 - \frac{x_2 - 3x_3\sqrt{x_6}}{x_2^2 + x_7x_1^3}}{ln\sqrt{x_9 x_{10}^2}} \tag{3}$$

where Y is the predicted output, used for process optimization, and the x variables are measured process parameters. The emulators are used for by-product optimization between two chemical plants in The Dow Chemical Company since March 2003.

6.3   Accelerated First-Principles Model Building

The key creative process in fundamental model building is hypothesis search. Unfortunately, the effectiveness of hypothesis search depends very strongly on creativity, experience, and imagination of the model developers. The broader the assumption space (i.e., the higher the complexity and dimensionality of the problem), the larger the differences in modelers' performance and the higher the probability for ineffective fundamental model building. In order to improve the efficiency of hypothesis search and to make the fundamental model discovery process more consistent, a new accelerated fundamental model building sequence is proposed. The key idea is to reduce the fundamental hypothesis search space by using symbolic regression, generated by GP. The main steps in the proposed methodology are shown in Fig. 11-1.

The key difference from the classical modeling sequence is in running simulated evolutions before beginning the fundamental model building. As a result of the GP-generated symbolic regression, the modeler can identify the key variables and assess the physical meaning of their presence/absence. Another significant side effect from the simulated evolution is the analysis of the key transforms with high fitness that persist during the GP run. Very often some of the transforms have direct physical interpretation that can lead to better process understanding at the very early phases of fundamental model development. The key result from the GP-run, however, is the list of potential nonlinear empirical models in the form of symbolic regression equations. The expert may select and interpret several empirical solutions or repeat the GP-generated symbolic regression until an acceptable model is found. The fundamental model building step 5 is based either on a direct use of empirical models or on independently derived first principles models revealed by the results from the symbolic regression. In both cases, the effectiveness of the whole modeling sequence could be significantly improved.

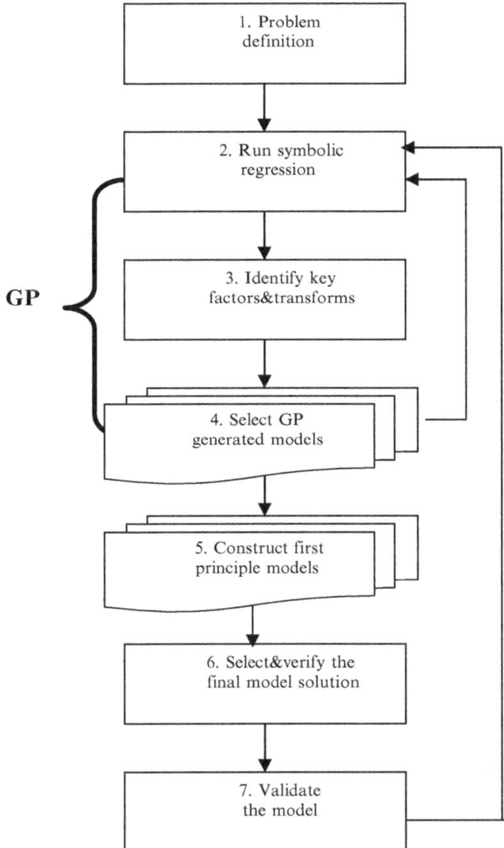

Figure 11-1. Accelerated new product development by using GP.

The large potential of genetic programming (GP)-based symbolic regression for accelerated fundamental model building was demonstrated in a case study for structure-property relationships (Kordon et al., 2002). The generated symbolic solution was similar to the fundamental model and was delivered with significantly less human efforts (10 hours vs. 3 months). By optimizing the capabilities for obtaining fast and reliable GP-generated functional solutions in combination with the fundamental modeling process, a real breakthrough in the speed of new product development can be achieved.

6.4   Linearized Transforms for Design of Experiments

GP-generated transforms of the input variables can eliminate significant lack of fit in linear regression models without the need to add expensive experiments to the original design, which can be time-consuming, costly, or maybe technically infeasible because of extreme experimental conditions. An example of such a type of application for a chemical process is given in (Castillo et al., 2002).

6.5   Complex Process Optimization

Process optimization is an area where EC technologies can make almost immediate economic impact and demonstrate value. Since the early 90s various evolutionary computation methods, mostly genetic algorithms, have been successfully applied in industry, including in The Dow Chemical Company. Recently, a new approach, Particle Swarm Optimization (PSO) (Engelbrecht, 2005) is found to be very useful for industrial applications. The main attractiveness of PSO is that it is fast, it can handle complex high-dimensional problems, it needs a small population size, and it is simple to implement and use. Different types of PSO have been explored in The Dow Chemical Company. For example, a hybrid PSO and Levenberg-Marquardt method was used for quick screening of complicated kinetic models (Katare et al., 2004). The PSO successfully identified the promising regions of parameter space that are then optimized locally.

6.6   Spectroscopy

Spectral data for measuring chemical concentrations consists of hundreds and even thousands of absorbance values per spectrum. Reducing this high dimensionality by appropriate variable selection is critical for developing reliable multivariate calibration models . A successful use of a genetic algorithm (GA), combined with PLS for the prediction of additive concentrations in a polymer films using Fourier transformed-infrared spectral data is presented in (Leardi et al., 2002). The variables selected by the GA are consistent with the expert

Table 11-1.   SELECTED EC APPLICATIONS IN DOW CHEMICAL

Application	Initial data size	Reduced data size	Model structure	Reference
Inferential sensors				
Interface level prediction	25 inputs x 6500 pts	2 inputs x 2000 pts	3 models 2 inputs	Kordon and Smits, 2001
Interface level prediction	28 inputs x 2850 pts	5 inputs x 2850 pts	One model 3 inputs	Kalos et al 2003
Emissions prediction	8 inputs x 251 pts	4 inputs x 34 pts	Two models 4 inputs	Kordon et al 2003b
Biomass prediction	10 inputs x 705 pts	10 inputs x 705 pts	9 models 2-3 inputs	Jordaan et al 2004
Propylene prediction	23 inputs x 6900 pts	7 inputs x 6900 pts	4 models 2-3 inputs	Jordaan et al 2004
Emulators				
Chemical reactor	10 inputs x 320 pts	10 inputs x 320 pts	5 models 8 inputs	Kordon et al 2003a
Accelerated modeling				
Structure property	5 inputs x 32 pts	5 inputs x 32 pts	One model 4 inputs	Kordon et al 2002
Structure property	9 inputs 24 pts	9 inputs 24 pts	7 models 3 -5 inputs	Kordon and Lue, 2004
Linearized transforms				
Chemical reactor model	4 inputs x 19 pts	4 inputs x 19 pts	3 transforms	Castillo et al 2002

knowledge and this could be the basis for automatic spectral analysis for these types of applications.

A selected set of EC applications from some of the above-mentioned application areas is given in Table 11-1. For each application the following information is given: initial size of the data set, reduced size of the data set (after variable selection and data condensation), model structure (number of inputs used in the selected final models and the number of models - some of them are used in an ensemble), and a corresponding reference with a detailed description of the application. In all the cases the final solutions were parsimonious models with a significantly reduced number of inputs.

7.   SUMMARY

EC has created significant value in the chemical industry by improving manufacturing processes and accelerating new product discovery. The key competitive advantages of EC, based on industrial applications in the company are defined as: no a priori modeling assumptions, high quality empirical models, easy integration in existing industrial work processes, minimal training of the final user, and low total cost of development, deployment, and maintenance. An overview of the key technical, organizational, and political issues that need to be resolved for successful application of EC in industry is given in the chapter. Examples of successful application areas are: inferential sensors, empirical emulators of mechanistic models, accelerated new product development, complex process optimization, effective industrial design of experiments, and spectroscopy.

Acknowledgments

The author would like to acknowledge the contribution in the discussed industrial applications of the following researchers from The Dow Chemical Company: Flor Castillo, Elsa Jordaan, Mary Beth Seasholtz, Guido Smits, Alex Kalos, Leo Chiang, Randy Pell, and Mark Kotanchek from Evolved Analytics and Riccardo Leardi from the University of Genoa.

References

Banzhaf, Wolfgang, Nordin, Peter, Keller, Robert E., and Francone, Frank D. (1998). *Genetic Programming – An Introduction; On the Automatic Evolution of Computer Programs and its Applications*. Morgan Kaufmann, San Francisco, CA, USA.

Bharadwaj, Sundar G, Varadarajan, P. Rajan, and Fahy, John (1993). Sustainable competitive advantage in service industries: A conceptual model and research propositions. *Journal of Marketing*, 57(10):83–99.

Castillo, Flor, Kordon, Arthur, Sweeney, Jeff, and Zirk, Wayne (2004). Using genetic programming in industrial statistical model building. In O'Reilly, Una-May, Yu, Tina, Riolo, Rick L., and Worzel, Bill, editors, *Genetic Programming Theory and Practice II*, pages 31–48. Springer.

Castillo, Flor A., Marshall, Ken A., Green, James L., and Kordon, Arthur K. (2002). Symbolic regression in design of experiments: A case study with linearizing transformations. In *GECCO 2002: Proceedings of the Genetic and Evolutionary Computation Conference*, pages 1043–1047. Morgan Kaufmann Publishers.

Christensen, Clayton M., Anthony, Scott D., and Roth, Erik A. (2004). *Seeing What's Next*. Harvard Business School Press, Boston, MA.

Engelbrecht, A. (2005). *Fundamentals of Computational Swarm Intelligence,*. Wiley, Chichester, UK.

Breyfogel III, F. (2003). *Implementing Six Sigma*. Wiley, Hoboken, NJ, 2nd edition.

Jain, L. and Martin, N., editors (1999). *Fusion of Neural Networks, Fuzzy Sets, and Genetic Algorithms: Industrial Applications*. CRC Press, Boca Raton, FL.

Jordaan, Elsa, Kordon, Arthur, Chiang, Leo, and Smits, Guido (2004). Robust inferential sensors based on ensemble of predictors generated by genetic programming. In Yao, Xin, Burke, Edmund, Lozano, Jose A., Smith, Jim, Merelo-Guervós, Juan J., Bullinaria, John A., Rowe, Jonathan, Kabán, Peter Tiño Ata, and Schwefel, Hans-Paul, editors, *Parallel Problem Solving from Nature - PPSN VIII*, volume 3242 of *LNCS*, pages 522–531. Springer-Verlag.

Katare, S., Kalos, A., and West, D. (2004). A hybrid swarm optimizer for efficient parameter estimation. In *Proceedings of Congress of Evolutionary Computation*, pages 309–315.

Kordon, A., Kalos, A., and Adams, B. (2003a). Empirical emulators for process monitoring and optimization. In *Proceedings of the IEEE 11th Conference on Control and Automation MED'2003*.

Kordon, A., Kalos, A., and Smits, G. (2001). Real time hybrid intelligent systems for automating operating discipline in manufacturing. In *Artificial Intelligence in Manufacturing Workshop Proceedings of the 17th International Joint Conference on Artificial Intelligence IJCAI-2001*, pages 81–87.

Kordon, A., Smits, G., Kalos, A., and Jordaan, E. (2003b). Robust soft sensor development using genetic programming. In *Nature-Inspired Methods in Chemometrics*. Elsevier, Amsterdam.

Kordon, Arthur, Castillo, Flor, Smits, Guido, and Kotanchek, Mark (2005). Application issues of genetic programming in industry. In Yu, Tina, Riolo, Rick L., and Worzel, Bill, editors, *Genetic Programming Theory and Practice III*, volume 9 of *Genetic Programming*, chapter 16, pages 241–258. Springer.

Kordon, Arthur, Pham, Hoang, Bosnyak, Clive, Kotanchek, Mark, and Smits, Guido (2002). Accelerating industrial fundamental model building with symbolic regression: A case study with structure-property relationships. In Davis, Lawrence "Dave" and Roy, Rajkumar, editors, *GECCO-2002 Presentations in the Evolutionary Computation in Industry Track*, pages 111–116.

Kordon, Arthur K. and Smits, Guido F. (2001). Soft sensor development using genetic programming. In Spector, Lee, Goodman, Erik D., Wu, Annie, Langdon, W. B., Voigt, Hans-Michael, Gen, Mitsuo, Sen, Sandip, Dorigo, Marco, Pezeshk, Shahram, Garzon, Max H., and Burke, Edmund, editors, *Proceedings of the Genetic and Evolutionary Computation Conference (GECCO-2001)*, pages 1346–1351. Morgan Kaufmann.

Kotanchek, Mark, Kordon, Arthur, Smits, Guido, Castillo, Flor, Pell, R., Seasholtz, M. B., Chiang, L., Margl, P., Mercure, P. K., and Kalos, A. (2002). Evolutionary computing in Dow Chemical. In Davis, Lawrence "Dave" and Roy, Rajkumar, editors, *GECCO-2002 Presentations in the Evolutionary Computation in Industry Track*, pages 101–110, New York, New York.

Leardi, R., Seasholtz, M. B., and Pell, R. (2002). Variable selection for multivariate calibration using a genetic algorithm: Prediction of additive concentrations in polymer films from fourier transforms-infrared spectral data. *Analytica Chimica Acta*, 461:522–531.

Parmee, I. (2001). *Evolutionary and Adaptive Computing in Engineering Design*. Springer, London, UK.

Smits, Guido and Kotanchek, Mark (2004). Pareto-front exploitation in symbolic regression. In O'Reilly, Una-May, Yu, Tina, Riolo, Rick L., and Worzel, Bill, editors, *Genetic Programming Theory and Practice II*, chapter 17, pages 283–299. Springer.

Chapter 12

TECHNOLOGY TRANSFER: ACADEMIA TO INDUSTRY

Rajkumar Roy and Jorn Mehnen

Decision Engineering Centre, Cranfield University, Cranfield, Bedfordshire MK43 0AL, UK

Abstract   High quality and innovation are major selling points in the technology market. Continuous improvement of products and the introduction of completely new products are a day to day challenge that industry has to face to keep competitive in a dynamic market. Customers desire changes when new materials and technologies become available. Consequently, new production views such as the whole life cycle cost of a product become an issue in industry. Keeping up with these changes is difficult and the application of the most recent technologies in a sound and effective way is often not straight forward. Academia is one of the sources of novel and scientifically well founded technologies. Furthermore, academia has a rich pool of thoroughly tested methods, well educated students and professional academics to deliver these methods. Technology transfer between academia and industry, therefore, is a productive way to bridge the gap between 'mysterious' theory and 'plain' practice. Various aspects of this transfer are discussed in this chapter. The most recent technology of multi-objective optimization is introduced to illustrate the challenges that come along with the cooperation between academia and industry.

Keywords: Knowledge Transfer, Modelling, Optimisation, Evolutionary Computing

1.   INTRODUCTION

The demand of industry to produce highly cost effective first-class goods implies that the production process has to be well understood and tuned to be as efficient as possible. Today, the application of computer based methods for estimating production cost and quality is well accepted. In many larger companies, it is realised that the virtual factory is a global major aim. Nevertheless, there is little work done in mathematical formulation and optimisation of these processes. A current survey shows that about 60 percent of the companies in automotive, aerospace, and steel do not have a systematic way to

R. Roy and J. Mehnen: *Technology Transfer: Academia to Industry*, Studies in Computational Intelligence (SCI) **88**, 263–281 (2008)
www.springerlink.com

develop process modeling. Much of the recent research in the area of process optimisation has dealt with either selection of a process model from a set of alternatives or uses a simple single-objective formulation that does not address the strong synergistic/anti-synergistic effects among individual activities. Current attempts suffer from serious limitations in dealing with the scalability requirements such as multi-objectivity and the complexity of real-life processes, including decision points or feedback loops. Therefore, there is a lack of technology in current literature for analysis and optimisation of computer-assisted real-world processes.

Solving difficult and challenging real-world problems need the collaboration of experts from various fields. Industry has a high demand for know-how that academia can provide. There exist highly sophisticated methods of approaching and solving technical problems that are well tried and tested. Surveys show that in many cases industry uses conventional techniques such as classical design of experiments in combination with finite elements method or CAE. Due to the fact that industry problems are very complex, these methods will not be sufficient in general to utilise the full potential of improvement of the processes. Much effort is still spent in manual optimisation of processes. Although manual optimisation has its rights, using powerful modelling and process optimisation techniques helps building a deeper knowledge and getting better results. Modern modelling and optimisation techniques which are carried out systematically do not necessarily need much time from the initial start to the delivery of practical solutions.

1.1 Technology Transfer: Industry And Academia

Transferring a technology into a new environment means that this technology should be understandable and maintainable over a long time. For industrial applications this implies training, consultancy and continuous updating of the system. A well planned collaboration between industry and academia takes these aspects into account. Industry can benefit over a long time from successful projects by having the chance to employ young and well educated personnel that are familiar with the environment. This allows exploiting proven scientific approaches to determine the best possible response to all kinds of challenges. Having powerful methods at hand and the right and educated personnel can give strong advantage on the daily market with significant improvements in performance and profits.

Academia provides best educated people and knowledge about modelling, simulation and optimisation techniques. The methods have a high degree of flexibility and can by adapt to various kinds of applications. Industry has to fulfill high day to day demands. This restricts the availability in exploring new market niches or major global trends. Education and training is important to support people in their daily work. Meanwhile, interaction with clever students helps industry workers being more creativity and becoming more open minded to new ideas.

A practical approach for interaction between industry and academia can be manifold. The structure of engagement varies depending on the needs and the nature of the projects being considered. A typical technology transfer may proceed in this or a similar way:

- Examine thoroughly the processes under consideration

- Conduct personal interviews with individuals

- Develop a process model and choose appropriate optimisation techniques

- Develop recommendations for improving the processes

- Take high level recommendations and expert knowledge into account

- Improve process model and solutions

- Present recommendations to stakeholder in the client organization

Involving expert's knowledge means strong interactions between academia and industry. As much input as possible from various people should be utilized to find perfectly tailored ready to use solutions. A key issue in solving any real-world problem is communication, understanding and trust. The relationship between industry and academia can be very fragile. The following is list of suggestions on how to make collaborations work better:

- Identify real needs on different time scales

- Have clear agreement on expectations

- Develop brief and decide together

- Select solutions and people cooperatively

- Keep regular review meetings and small deliverables on regular basis

- Be reasonably flexible to respond to needs

- Bring in other expertise if necessary

- Maintain the relationship between project partners

- Support independency and trust by training and education

- Have a long term strategy

Identifying the real needs is a relevant factor that defines the complexity of a solution approach. Here again, the first step is to get clear agreements of the expectations between industry and academia. Finding a common language helps avoid misunderstandings. For example, graphs and simplifications may improve understanding. Also, a step by step approach helps create trust, reduce disturbances and improves knowledge.

Modern optimisation tools can be understood on a very basic level. This means that the time consuming training of staff is not necessary. The optimisation methods are quite intuitive because many algorithms are motivated by well known natural principles such as evolutionary processes. Although these algorithms are highly complex, the application of these methods does not require a deep understanding of the methods. The same is true for many soft computing techniques which can be used for modelling. For example, fuzzy logic, which supports the usage of imprecise human language to describe process behaviour in a very precise manner. There exist several real-world applications based on soft computing techniques, which indicate these technique are very effective.

In Section 2, an overview of technologies for modelling and optimisation is given. It consists of two parts: process design and multi-objective optimization design. Section 3 discusses the integration of models and optimisation strategies in the real-world environment. It covers integration aspects such as visualisation and interfaces. Our concluding remarks are given in Section 4.

2. PROCESS DESIGN AND EVOLUTIONARY OPTIMISATION

2.1 Process Design Aspects

In many technical applications it is necessary to have a numerical description of a specific section of a real-world process in a computer. This implies a transfer or mapping of technological knowledge to a functional or relational description. This functions and relations are referred to as models in the rest of the chapter.

Processes knowledge can either be given explicitly or implicitly. Examples of implicit descriptions are observed data or experience of experts. The latter is sometimes hard to formalise. Explicit formulation can be avoided by user interaction methods. Alternatively, soft computing techniques such as fuzzy logic or neural networks help getting explicit mathematical descriptions of expert knowledge. Fuzzy logic (FL) is a technique which uses linguistic terms to capture qualitative values as that used in spoken language (Oduguwa et al., 2007). Artificial Neural Networks (ANN) classify or approximate given data. Plain ANNs generally use crisp numerical values. Of course both techniques can be blended very well (Gorzalczany, 2002). It is common and recommended practice to pre-process data before feeding it to ANN or FL systems for model building. Pre-processing should extract certain features from the data and hence reduce the amount of noise. This improves processing speed and learning rate. An an analogy to real optical data processing, visual images are not directly fed to the brain but pre-processed on their way from eye to brain. Similar to biological brain, artificial neural networks are developed in a completely self-organised manner and can be very powerful in modeling systems.

Explicit knowledge is described in a formal mathematical way. Formal models either use discrete or continuous numerical values or both. The corresponding data may appear in 'atomic' numeric form. It may also be organised in more complex data structures such as graphs or tables. High level data structures use meta languages such as XML (Harold and Means, 2004) or CAD-formats such as IGES (US PRO, 1993) or STEP (STEP, 1994). Optimisation methods most often use numbers only. Genetic programming may be seen as an outlier in this context because it uses also higher level structures.

Typical examples of explicit models using discrete data are schedules or state transition graphs. Often discrete models have a combinatorial character and can be very hard to solve (Ehrgott, 2006; Murat and Paschos, 2006). Continuous quantitative numeric values are used in many physical or engineering models.

Explicit continuous formulas describe complex differential equations or arbitrary functional relations (Gershenfield, 1998). These formulations are the most common in engineering and natural sciences. Continuous models or at last partially continuous models are advantageous for optimisation. Some optimisation methods explicitly require derivations to calculate gradients (e.g. Sequential Quadratic Programming SQP (Boggs and Tollow, 1996)). But even for direct methods (e.g. Nonlinear Simplex Method (Nelder and Mead, 1965)) with known convergence properties (Lagarias et al., 1998) and stochastic optimisers, which do not need derivatives, a smooth fitness function with large attraction areas containing the optimum are preferable to 'rough' multimodal functions.

In practice it often happens that models have plane areas which do not give any hint on beneficial search directions. In this case stochastic optimisers show Brownian movements and hill climbing optimisers may stop moving completely. Constraints with infinite penalty often show this characteristic and, therefore, fitness functions do not show grident information.

A similarly difficult problem is the needle in the haystack. Here a continuous or non-continuous function has a singular very local optimum. This case appears if most parameter combinations yield insufficient results and only one single combination produces outstanding good results.

Nonlinear problems may be very sensitive to small errors (e.g. Argyris et al., 1994). The decision to use complex nonlinear models can be very important because extrapolation and forecasting needs should be taken into account.

Modelling noisy input or output data can be used to describe uncertainty. Either noise can be seen as a disturbing effect or as a driving force. Models such as Markov chains, Gibbs Fields, Monte Carlo Simulations and Queues (Brmaud, 1999) or evolutionary algorithms (Bäck et al., 2000) make explicit use of stochastic variation. In this vein, statistical models have a great influence in process modelling and play a relevant role in systematic technology transfer. Statistical methods are generally well founded and can be applied very systematically and precisely with a wide range of practical applications (e.g. Khattree and Rao, 2003).

The number of input variables used in a model influences the model complexity. It is possible that many hundreds or thousands of parameters may often have an influence on a particular process. It may be possible to integrate all these parameters but most of the times this is not reasonable. Linear models often deal with thousands of input variables. Continuous nonlinear multimodal models as well as discrete models using many variables may suffer from the curse of dimensions, i.e. the solution of a problem becomes unacceptably inefficient.

Statistical Design of Experiments (DoE) methods provide techniques to pre-select the most relevant data (e.g. Francoise and Lavergne, 2001; Bartz-Beielstein 2006; Oduguwa et al. 2007). If evaluation of a function is very expensive and derivatives are not available, the Kriging method may be useful to create surrogate models. Here, linear approximations with nonlinear basis functions are used to form an interpolation model with known error limits. This model is used as a surrogate to find new potential solution candidates which are evaluated in real-world or with more expensive models. Complex real-world design optimisation problems have been solved this way at General Motors Detroit in 1997.

The number of output variables usually corresponds to the number of results a model generates. The results often directly reflect the qualities of a process or may be the basis for several quantitative values. Multi-objective models map several input values to several output values (Miettinen, 1998; Deb, 2001; Collette and Siarry, 2003; Ehrgott, 2006). Different from single-objective approaches, multi-objective application may have conflicting or contradictory objectives. For example, in engineering applications a significant reduction in the weight of a structure may be advantageous with respect to cost but may be disadvantage to stability. Here, both costs as well as stability are targets to be optimized. Because both targets are conflicting it is only possible to get trade-off solutions. Targeting multiple objective has a decisive influence on the optimisation results and hence the optimisation strategies to be applied. Multi-objective optimisation results are generally not single point solutions but rather a set of alternative solutions from which a decision maker can choose. This means for real-world applications that an engineer should have easy and fast access to human interpretable solutions. The idea of multi-objective modelling introduces a certain kind of uncertainty that opens up a range of alternative solutions. Industry users can benefit from this technique especially when combined with evolutionary methods (see below) because typically new and unforeseen solutions are generated. Alternative technical solutions introduce new challenges in manufacturing processes yielding to new and innovative products.

Introduction of expert knowledge into multi-objective models is a very important issue. Experience of engineers can be used to reduce the solution set to practically desirable subsets. Speaking in terms of costs, these subsets may represent affordable solutions. Expert knowledge can be introduced e.g. by using constraints. More efficient techniques that avoid constraints are also developed (Branke and Deb, 2005) or (Trautmann and Weihs 2006). The latter introduce so called desirability functions. This technique maps the model output to a desirability space where each output is assigned a desirability value. This method has been tested successfully in combination with evolutionary multi-objective optimisation in real-world environments (Mehnen and Trautmann, 2006).

Using constraints becomes more and more relevant in real-world applications. While well known in linear programming, constraints were often neglected in nonlinear programming community. Constraints or restrictions introduce limits to the feasible search space. Either input parameters or output parameters can be constrained. Constraints can be either linear or nonlinear. The modelling of constraints is not an easy task, especially, if many constraints are used. It may happen that either the constraints 'over restrict' the search space in a way that no feasible solution can be found or they limit each other in a way that restrictions become redundant. An evaluation of a constraint can be time consuming and generally constraints cause problems during optimisation. Typically, penalty functions are used to model constraints. Penalty functions introduce an exponential or polynomial additional malus to the quality of an infeasible solution. Optimisation under constraints increases the complexity of a problem. A practical approach to overcome problems caused by restrictions can be an annealing technique, where the penalties change in intensity from weak to strong over time. Another solution is to design a model in a way that all elements in the solution space are always feasible. If this is not possible, sensible 'repair' mechanism during evaluation or optimisation of the model may be applied. Repair strategies should be implemented carefully because they may introduce a bias to the search process. Other approaches interpret constraints as additional objectives. Further references on constraints can be found in (Michalewicz, 1995; Coello Coello, 1999; Mezura-Montes and Coello Coello, 2006; Chootinan and Chen, 2006; Luus et al., 2006).

It should be noted that in this text, model, solver/simulation and optimiser are considered to be basically different. A model is the description of an issue in real world while the solver or simulation evaluates the model. For example, Finite Element Method is usually identified as a model but actually it is the combination of a solver and a model. The model uses differential equations while the solver tries to find a solution of the differential equations. Simulations are solvers that drive e.g. a discrete model to switch between different states. Optimisers try to find best solutions with respect to a given problem. While a problem and its corresponding mathematical model may be simple, e.g. 'find the highest natural number', its evaluation may take a long, even infinite, time. The same can be true for optimisation where the evaluation of a problem may be efficient but finding an optimum may be hard. A typical problem of this class is the 'travelling salesperson problem'. Another example in a continuous domain is the problem of the 'needle in the haystack', where a single very discrete optimum has to be found under infinite many other possible solutions.

This means that choosing the right description of a problem has a strong influence on its algorithmic complexity and, therefore, should be considered with much care.

Introducing a model to practical applications needs thorough investigations and background knowledge about theoretical as well as practical aspects of a problem. In practice, the qualities of the data and the amount of knowledge available to form a model have a strong influence on the modelling process. In practice, of course, the time available to set up a model, the desired quality of the results, and the time available for solving a problem also influence the modelling process. Model design is a multi-objective task and generally trade-off solutions have to be found. Sometimes models using qualitative data or rough approximations are much more efficient or near to real life than highly sophisticated quantitative numeric evaluations. Other times, high precision models satisfy real-world demands, e.g. the estimation of safety tolerances or risk, better.

Modelling can become a complex task and requires a lot of experience. The choice of the correct model has a decisive influence on the run time and the quality of the solutions produced. Outsourcing the complex task of analysis and modelling to academia can be very valuable for industry. University experts approach problems with much experience and know several problem domains very well. They can introduce the most recent technologies to industry and hence provide competitive edge.

2.2   Multi-Objective Optimisation Design Aspects

Models, optimisation techniques and the technological environment have to match perfectly each other to get high quality and efficient solutions. If possible, modelling and optimisation should be harmonised during the design phase. Nevertheless, several decisions have to be faced when optimisation methods are to be applied.

One should be aware that there is no globally best optimiser for any problem. In particular all algorithms that search for an extremum of a cost function perform exactly the same, when averaged over all possible cost functions (Wolpert and Macready, 1995). This became known as the famous No Free Lunch (NFL) theorem. Of course, this statement is true for real-world problems as well. Nevertheless, in general not all problems have to be considered. Droste, Jansen and Wegener showed that for each function, which can be optimised efficiently, there are many related functions that cannot be solved efficiently using the same optimiser (Droste et al., 2002). This became known as the Almost No Free Lunch Theorem (ANFL). This means that search methods can be very successful if they utilise knowledge about a problem correctly.

The main knowledge about a problem is reflected by the model and generally only some domain knowledge is introduced into the optimiser to avoid biasing effects or getting stuck in suboptimal solutions. This means that the correct matching of a problem class with the optimiser influences the efficiency and quality of the optimisation process.

Exploration and exploitation are two key terms in optimisation. They indicate the tension between the desire to find new solutions and the demand to find solutions quick. This tension becomes very strong in practical applications where, on the one hand, innovative and even unexpected results should be produced while, on the other hand, good results should be available as fast as possible.

Deterministic search strategies perform very well on many problem classes and have a well elaborated and sound mathematical background. Sometimes the time and space complexity of a problem and the efficiency of the optimiser can be estimated precisely. This means that one can give upper or lower limits of the expected solution quality and the run time. In lucky cases, where problem class and the corresponding efficient optimiser are known explicitly, it is obvious to use deterministic strategies.

In many practical applications the problem class is not known or the design of the model is not explicitly given or the model is hidden. This may be due to e.g. industry confidentiality policy or simply because the problem is much too complex to be understood completely by human beings. In this case two possible approaches are common: surrogate modelling, i.e. simplifying the problem using e.g. DoE approaches (see above) or applying general purpose solver such as stochastic meta-heuristics. Evolutionary algorithms, ant colony optimisation or particle swarm optimisation are examples for stochastic meta-heuristics.

The term Evolutionary Algorithms (EA) covers a whole family of computation intelligence (CI) methods such as Genetic Algorithms (GA) (e.g. Goldberg 1989), Evolution Strategy (ES) (e.g. Schwefel 1995), Evolutionary Programming (EP) (e.g. Fogel 1991), and Genetic Programming (GP) (e.g. Koza 1992). EA are stochastically driven algorithms which mainly consists of a set of solution candidates - called population - which undergo a continuous multiplication, variation and selection process in order to find solutions, which suit the optimisation problem best. Today, EA are widely accepted in industry and belong to the state-of-the-art in nonlinear optimisation. Relevant improvements of one or the other technique can still be registered. An example is the Covariance Matrix Adaptation-ES (CMA-ES). (Hansen and Ostermeier 2001).

In a multi-objective context, mathematical background theory (Miettinen 1998, Ehrgott 2005) on deterministic optimisation strategies is rare. This is because in general multi-objectiveness is mathematically very difficult to solve. One way out of this dilemma is mapping the multi-objective problem to a single-objective problem and to apply adequate well known single-objective optimisers (Collette and Siarry 2003, Ehrgott 2005, T'kindt and J.-Ch. Billaut 2006). In practice this may work very well and fast. Unfortunately, scalarisation techniques are only applicable when the desired area of the trade-off solution is known in advance. Examples of scalar optimisation techniques in a multi-objective continuous domain are:

- Weighted linear aggregation of fitness values (disc. in Ehrgott 2005)

- Goal programming and Goal attainment (disc. in Collette 2003)

- Method of Hillermeier (Hillermeier 2001)

- Method of Fliege (Fliege 2001)

- Keeney-Raiffa aggregation method (Keeney-Raiffa 1993)

- Method of antiparallel lines (Ester 1987)

- Rotated weights metrics method (Zeleny 1982)

- Lin-Tabak-Algorithm (Tabak 1979)

- Lin-Gisey-Algorithm (Giesy 1978)

- Strict-equality-constraints method (Lin 1976)

- μ-constrainted method (HLW 71)

The more difficult a real-world problem is, the less knowledge about it is available. This means, in the extreme case, that neither knowledge about the class of the problem is given nor the structure of the possible alternative multi-objective solution set can be provided. In this case only clever guessing is possible.

Multi-objective evolutionary algorithms (MOEA) must be seen as generally different from single-objective EA. Most MOEA (Multi-Objective Evolutionary Algorithms) generate Pareto-sets and Pareto-fronts which reflect the set of best compromises and their respective fitness values (Deb 2001). The selection of 'one best' individual from a population during selection is not unique any more. Objectives like the spread of the generated Pareto-front-approximation and the distance from the real Pareto-front have to be taken into account. Since the comparison of the solution sets is not straight forward or unique since the estimation of the convergence speed of a MOEA becomes a difficult task. Nevertheless, there are some measuring methods (in the MOEA context called metrics, although often not being a real metric in a strict mathematical sense) which can be applied for estimating the quality of a Pareto solution (CoelloCoello and VanVeldhuizen 2002, Colette and Siarry 2003).

MOEA are applied to multiple fitness functions which map homogenous n-dimensional decision spaces to m-dimensional (m>1) objective spaces. In many practical cases both spaces are real-valued. The selected list of the most famous examples of this class of Pareto-based MOEA are (ordered by year of appearance):

- SMS-EMOA: S-Metric Selection Evolutionary Multi-Objective

- Algorithm (Emmerich et al. 2005)

- Epsilon-MOEA (Deb et al. 2003)

- SPEA2: Strength Pareto Evolutionary Algorithm 2 (Zitzler et al. 2002)

- MOMGA-II: Multi-Objective Messy Genetic Algorithms (Zydallis et al. 2001)

- PESA: Pareto Envelope-based Selection Algorithm (Corne et al. 2000)

- PAES: Pareto Archived Evolution Strategy (Knowles et al. 1999)

- NSGA-II: Non-dominated Sorting Genetic Algorithm (Deb 1993)

- MOGA: Multi-Objective Genetic Algorithm (Fonseca 1993)

- etc.

More than forty other different multi-objective evolutionary algorithms and many more variants could be listed. This indicates the fact that there is still a big gap between theory and practice in MOEA. A lot of specific adaptations of the standard algorithms have been 'mushrooming up' for the last seven years. Currently, a systematic mathematically founded technology transfer which may lead to a more efficient implementation of MOEA in real-world applications does not exist.

As an interesting fact, it should be mentioned that only few multi-objective algorithms are capable of handling single as well as multi-objective problems. Examples are the Prey-Predator-MOEA (Schmitt 2005) and a specific MOEA developed by (Deb and Tiwari 2005). The integration of expert knowledge into MOEA leads to advanced strategies. Expert knowledge can help to limit the solution space with respect to size, complexity or quality. In industrial practice, it is often that a small improvement could lead a large amount of financial rewards. Hence, it may not be necessary to search for the very best solution. Algorithms such as the epsilon-MOEA (Deb 2003) already account for that.

A designer of an optimisation program who is examining the problem often gets more and more detailed knowledge about a problem. Integration of expert knowledge can be handled by developing MOEA-hybrid solutions. In this case standard MOEA are blended with deterministic or other stochastic optimisers. These modifications have major impact on the design and performance of the MOEA. Changing an EA always implies the danger that experience about the algorithm's general behaviour and reaction on parameter variations gets lost. Therefore, implementation of hybrids should be done with much care.

Three possible approaches for generating hybrid EA/MOEA can be identified: blending on population level, blending on time scale or blending on operator level. Introducing deterministic algorithms can be done on population level by either using the EA exploration features to find new starting points for an exploitive deterministic search or using the EA's explorative and exploitive strength to search deterministically after several iterations of the EA cycle. Switching completely from EA to deterministic search or vice versa after several search steps are two extreme examples of blending on the time scale. Introducing knowledge into the operators can be very tricky because the EA should generally be able to have best exploration abilities, which may be violated strongly by determinism. The problems of 'when to switch' and 'what type of deterministic algorithm should be used and where' depend on the optimisation task at hand. In practice, only experience and discussions between academia and industry and, in the worst case, just clever trial and error techniques are available.

Models, as well as optimisation algorithm tuning, verification and comparison are important topics. Tuning of models and optimisers is often a continuously ongoing process.

Evolutionary algorithms especially come up with slightly different solutions and statistical analysis of the results become mandatory. In practice, of course, only the best results are of particular interest. Visualisation and comparison of the performance of MOEA is more complex than for single-objective algorithms because multiple differences between whole sets have to be measured. Again Pareto-metrics can be used to compare the results of different MOEA.

A sound and fair comparison of optimisers can only be done if the same models and starting conditions are applied. Furthermore, it is important to use the best parameter settings of each optimiser for the comparison. Tuning the parameter settings of the optimisers means optimising the optimisation algorithm. This step can be quite expensive. DACE (Design and Analysis of Computer Experiments) can help reducing the amount of experiments while keeping expressiveness of the results at a high level (e.g. Bartz-Beielstein 2006).

Many MOEA generate solutions that are presented to a decision maker in an a posteriori approach. Looking at the Pareto front helps selecting solutions of highest interest to the user. Aspects such as e.g. cost or quality can be balanced this way easily. Picking points from the Pareto front, identifying these points with solutions in the Pareto set and visualising these solutions can be very helpful in practical applications where each single solution can be associated to a realisation in real world. The visualisation of e.g. geometric shapes of objects provides a fast and intuitive approach for engineers (Mehnen 2005). This allows a direct interaction with the optimiser and may also provide hints for further tuning of the underlying model. Low dimensional visualisation of e.g. three dimensional objects which are optimised using more than three objectives is one way to interpret solutions found in higher dimensional objective spaces. In order to keep efficiency of the optimiser and interpretability reasonable high, in practice generally objective spaces with more than seven objectives are quite seldom (Coello Coello and Van Veldhuizen 2002).

3. PRACTICAL INTEGRATION INTO REAL-WORLD ENVIRONMENTS

3.1 Interfaces

Implementing a new simulation and optimisation tool for practical applications always implies considerations about the environment it will be used in. Here, the term 'environment' stands for software, hardware and workshop environment. Considering the interfaces between the system and its environment can be crucial for the system's practical applicability. Having standardised data exchange formats (e.g. XML, STEP, CAD-formats, human readable ASCII tables etc.) can help reducing a lot of extra work and makes the system more interpretable.

Figure 12-1 gives an overview on the interaction between model, optimisation system, human and the environment. Interfaces have to be able to handle various types and structures of input and output data to connect the model with the optimisation tool. Also, fast interfaces are necessary for grid networking (external calculation) or human interaction.

3.2 Visualisation

Visualisation is an important technique to show results in an intuitive and compact manner. In multi-objective optimisation the visualisation of the

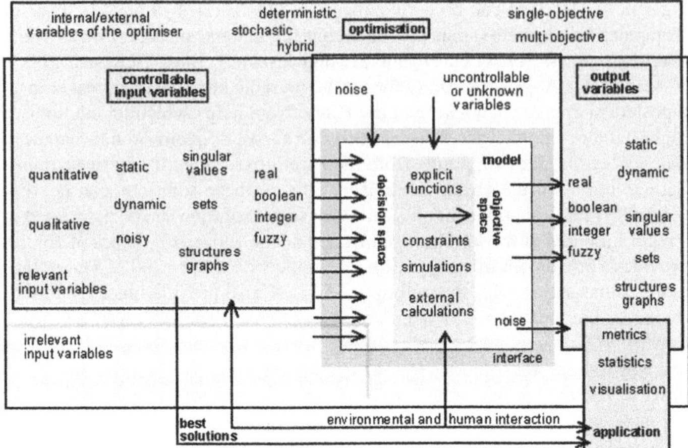

Figure 12-1.   Interaction between model, optimisation and environment.

Pareto-fronts and Pareto-sets is necessary for the user to select the solutions that solves the problem that best.

Pareto fronts can be displayed easily as point clouds in two or three dimensional space. This is the most commonly used presentation method. Points in higher dimensional spaces can be shown using scatter plots, where two or three out of m dimensions are displayed in one graph. If m is high, the number of plots can be large $(m!/(4(m-2)!))$.

Another technique is called 'Andrew waves' (Andrews 1972), which is similar to profile plots where each objective is modelled by a polyline or a trigonometric polygon, respectively. These techniques easily confuse the user if the number of objectives is large.

Star and sun plots can display m-dimensional single multivariate data. In these plots, the lengths of the rays are proportional to the value of fitness value. Another ploting system is Glyphen, which follows the idea of star or sun plots (Rinne 2003). An industrial application of an interactive star plot for use on internet also exists. The system NAVIG2 allows choosing different GPS navigator systems from a database according to the desires of the client.

In many real-world applications two or three dimensional phenotypes of the corresponding high dimensional genotype representations can be used to visualise Pareto solutions. It can be very helpful in discussions to show selected items on the Pareto front in combination with their corresponding phenotypes (Mehnen et al., 2007). Moreover, virtual reality environments are especially helpful to get intuitive impressions.

4. CONCLUSIONS

Introducing new techniques into real-world environments in a sound and efficient way demands an expert approach. Technology transfer between academia and industry is a sensible matter and should be done with much care. The application of modern modelling, simulation and optimisation methods is not straight forward. Challenging real-world cases have to be treated in their own and very special way and often these challenges lead to new ways of solving problems and, hence, to new knowledge.

Multi-objective evolutionary optimisation is an example where new and unusual concepts are used. As shown in this chapter, models and optimisation methods should be designed to match each other perfectly. Utilising these methods with their full power implies keeping the whole process in mind, i.e. from the first interview in industry to the final product. This needs much time, effort, expert knowledge and consultancy. Technology transfer can lead to new technologies as well as to financial benefits, in addition to more highly educated people with many promising long-term perspectives.

References

D.F. Andrews (1972) Plots of High-Dimensional Data *Biometrika* 28:125-136

J. Argyris and G. Faust, and M. Haasse (1994) *An Exploration of Chaos.* North-Holland Publishing, NL.

Th. Bäck, D. B. Fogel, and T. Michalewicz (2000) *Evolutionary Computation 1 and 2,* Institute of Physical Publishing IoP, Bristol, UK

Th. Bartz-Beielstein (2006) *Experimental Research in Evolutionary Computation: The New Experimentalism* Natural Computing Series, Springer, Berlin.

P.T. Boggs, T.W. Tollw (1996) Sequential Quadratic Programming. *Acta Numerica*, 4,1-51

J. Branke and K. Deb. (2005) Integrating User Preferences into Evolutionary Multi-Objective Optimization. In Y. Jin (editor), *Knowledge Incorporation in Evolutionary Computation*, Springer, pp. 461-477, Berlin Heidelberg.

P. Brmaud (1999) *Markov Chains, Gibbs Fields, Monte Carlo Simulation, and Queues.* Springer, New York

P. Chootinan and A. Chen (2006) Constraint Handling In Genetic Algorithms Using A Gradient-Based Repair Method. *Computers and Operations Reseach.* 33(8):2263-2281.

C. Coello Coello (1999) A Survey of Constraint Handling Techniques used with Evolutionary Algorithms. *Technical Report Lania-RI-9904, Laboratorio, Nacional de Informtica Avanzada.*

Y. Collette and P. Siarry (2003) *Multiobjective Optimization: Principles and Case Studies.* Decision Engineering Series. Springer, Berlin.

D.W. Corne, J.D. Knowles, and M.J. Oates (2000) The Pareto Envelope–based Selection Algorithm for Multiobjective Optimization. *Proceedings of the Parallel Problem Solving from Nature VI*, M. Schoenauer et al. (eds.), Springer, Berlin, pp. 839-848.

K. Deb (2001) *Multi-Objective Optimization Using Evolutionary Algorithms.* John Wiley, Chichester, UK

K. Deb, M. Mohan, M. and A. Mishra (2003). A Fast Multi-objective Evolutionary Algorithm for Finding Well-Spread Pareto-Optimal Solutions. KanGAL Report No. 2003002.

K. Deb, S. Agrawal, A. Pratab, and T. Meyarivan (2000) A Fast Elitist Non-Dominated Sorting Genetic Algorithm for Multi-Objective Optimization: NSGA-II. *Proceedings of the Parallel Problem Solving from Nature VI Conference*, Springer. Lecture Notes in Computer Science, No. 1917, Paris, France, M. Schoenauer et al. (eds.), pp. 849-858.

K. Deb and S. Tiwari (2005) Omni-optimizer: A Procedure for Single and Multi-objective Optimization. *Evolutionary Multi-Criterion Optimization. Third International Conference*, EMO 2005. C. A. Coello Coello et al. (eds.), pp. 47-61, Springer. Lecture Notes in Computer Science Vol. 3410.

S. Droste, Th. Jansen, and I. Wegener (2002) Optimization with randomized search heuristics - the (A)NFL theorem, realistic scenarios, and difficult functions, *Theoretical Computer Science*, 287, pp. 131-144.

M. Ehrgott (2005) *Multicriteria Optimization.* 2nd edition, Springer, Berlin.

M. Emmerich, N. Beume, and B. Naujoks (2005). Multi-objective optimisation using S-metric selection: Application to three-dimensional solution spaces. In B. McKay et al., Eds., *Proc. of the 2005 Congress on Evolutionary Computation (CEC 2005)*, Edinburgh, Band 2, pp. 1282-1289. IEEE Press, Piscataway, NJ, USA.

J. Ester (1987) *Systemanalyse und mehrkriterielle Entscheidung.* VEB Verlag Technik, Vol. 1, Berlin, Germany.

J. Fliege (2001) *Approximation Techniques for the Set of Efficient Points.* Habilitation, Dortmund, Germany.

D. B. Fogel (1991) *System Identification through Simulated Evolution: A Machine Learning Approach to Modeling.* Ginn Press, Needham Heights, MA, USA.

C.M. Fonseca and P.J. Fleming (1993). Multiobjective Genetic Algorithms. In *IEE Colloquium on Genetic Algorithms for Control Systems Engineering*, pp. 6/1–6/5, IEE, UK.

O. Francois and C. Lavergne (2001) Design of evolutionary algorithms - A statistical perspective. *IEEE Transactions on Evolutionary Computation,* Vol. 5, No. 2. pp. 129-148.

N. Gerschenfield (1998) *The Nature of Mathematical Modeling.* Cambridge Univ. Press, UK

D.P. Giesy (1978) Calculations of Pareto Optimal Solutions to Multiple Objective Problems using Threshold of Acceptance Constraints. *IEEE Transactions on Automatic Control*, AC-23(6):1114-1115.

D. E. Goldberg (1989) *Genetic Algorithms in Search, Optimization and Machine Learning.* Addison Wesley, Reading, USA.

M. B. Gorzalczany (2002) *Computational Intelligence Systems and Applications: Neuro-fuzzy and Fuzzy Neural Synergisms.* Studies in Fuzziness & Soft Computing, Physica, Heidelberg, Germany.

Y.Y. Haimes, L.S. Lasdon, and D.A. Wismer (1971) On a bicriterion formulation of the problems of integrated system identification and system optimization. *IEEE Transactions on Systems, Man, and Cybernetics*, Vol. 1, pp. 296-297.

E.R. Harold, W.S. Means (2004), *XML in a Nutshell,* O'Reilly

N. Hansen and A. Ostermeier (2001). Completely Derandomized Self Adaptation in Evolution Strategies. *Evolutionary Computation*, 9(2), pp. 159-195.

C. Hillermeier (2001) Nonlinear multiobjective optimization: a generalized homotopy approach. Birkhuser Verlag, 135, *International series of numerical mathematics*, Basel, Switzerland.

F. Jarre and J. Stoer (2003) *Optimierung*, Springer, Berlin.

R.L. Keeney and H. Raiffa (1993) *Decisions with Multiple Objectives: Preferences and Value Tradeoff.* Cambridge University Press, Cambridge, UK.

R. Khattree and C.R. Rao (2003) *Handbook of Statistics.* Vol. 22, North-Holland Pub., NL

J.D. Knowles and D.W. Corne (1999) The Pareto Archived Evolution Strategy: A New Baseline Algorithm for Pareto Multiobjective Optimisation. In *Proceedings of the 1999 Congress on Evolutionary Computation (CEC'99)*, pp. 98-105

J. Koza (1992) *Genetic Programming: On the Programming of Computers by Means of Natural Selection.* MIT Press, USA.

J.C. Lagarias, J.A. Reeds, M.H. Wright, P.E. Wright (1998) Convergence properties of the Nelder-Mead simplex method in low dimensions. *SIAM J. Opt.* 9(1):112-147

J.G. Lin (1976) Multipleobjective problems: Paretooptimal solutions by proper equality constraints. *IEEE Transactions on Automatic Control,* 5(AC21):641-650.

R. Luus and K. Sabaliauskas, and I. Harapyn (2006). Handling Inequality Constraints in Direct Search Optimization. *Engineering Optimization*, 38(4):391-405

J. Mehnen (2005) M *ehrkriterielle Optimierverfahren fr produktionstechnische Prozesse*. Habilitationsschrift, Universitt Dortmund, Vulkan Verlag, Essen, Germany.

J. Mehnen, Th. Michelitsch, and C. Witt (2007) Collaborative Research Centre 531: Computational Intelligence - Theory and Practice. Oldenbourg Wissenschaftsverlag, Munich, Germany, *it - Information Technology*. 49(1): 49-57.

J. Mehnen, H. Trautmann (2006) Integration of Expert's Preferences in Pareto Optimization by Desirability Function Techniques. In: *Proceedings of the 5th CIRP International Seminar on Intelligent Computation in Manufacturing Engineering (CIRP ICME '06)*, Ischia, Italy, R. Teti (ed.), pp. 293-298, Copyright C.O.C. Com. org. Conv.

E. Mezura-Montes, C. Coello Coello (2006) A Survey of Constraint-Handling Techniques Based on Evolutionary Multiobjective Optimization. *Workshop paper at PPSN 2006*, Iceland.

Z. Michalewicz (1995) A survey of constraint handling techniques in evolutionary computation methods, *Proc. of the 4th Annual Conf. on Evolutionary Programming*, MIT Press, Cambridge, MA, J. R. McDonnell, R. G. Reynolds, and D. B. Fogel (Eds), pp. 135-155.

K. Miettinen (1998) *Nonlinear Multiobjective Optimization*. International Series in Operations Research & Management Science, Kluwer Academic Publishers, Dordrecht, NL.

C. Murat and V. Th. Paschos (2006) *Probabilistic Combinatorial Optimization on Graphs*. ISTE Publishing Company, Washington, DC, USA.

J.A. Nelder and R. Mead (1965) A simplex method for function minimization, *Comp. Journal*, 7:308-313

V. Oduguwa, R. Roy and D. Farrugia (2007) Development of a soft computing based framework for engineering design optimisation with quantitative and qualitative search spaces. *Applied Soft Computing*, 7(1):166-188.

H. Rinne (2003) *Taschnenbuch der Statistik*, Harri Deutsch Verlag

K. Schmitt, J. Mehnen, and Thomas Michelitsch (2005). Using Predators and Preys in Evolution Strategies, in Hans-Georg Beyer et al. (editors), *2005 Genetic and Evolutionary Computation Conference (GECCO'2005)* , pp. 827-828, Vol. 1, ACM Press, NY, USA

H.-P. Schwefel (1995) *Evolution and Optimum Seeking*. Wiley-Interscience, NY, USA.

STEP International Organization for Standardization (1994) ISO 10303-42, *Int'l. Organization for Standardization*, Vol. 42, Switzerland

D. Tabak (1979) Computer Based Experimentation with Multicriteria Optimization Problems. *IEEE Transactions on Systems, Man and Cybernetics*, SMC9 (10):676-679.

V. T'kindt and J.-Ch. Billaut (2006) *Multicriteria Scheduling: Theory, Models and Algorithms*. Springer, Berlin.

H., Trautmann, C. Weihs (2006) On the distribution of the desirability function index using Harrington's desirability function. *Metrika*, 62(2):207-213.

US PRO (1993) IGES 5.2 An American National Standard, ANS US PRO-IPO-100-1993, U.S. Product Data Association, Gaithersburg, MD, USA

D. H. Wolpert, and W. G. Macready (1995) No Free Lunch Theorems for Search. *Technical Report SFI-TR-95-02-010*. Sante Fe, NM, USA: Santa Fe Institute.

M. Zeleny (1982) *Multiple Criteria Decision Making*. McGraw-Hill, New York, USA.

E. Zitzler, M. Laumanns, and L. Thiele (2001) SPEA2: Improving the Strength Pareto Evolutionary Algorithm. *Technical Report No. 103,* Computer Engineering and Communication Networks Lab (TIK), ETH Zrich, Switzerland.

J.B. Zydallis, D.A. van Veldhuizen, and G. Lamont (2001) A Statistical Comparison of Multiobjective Evolutionary Algorithms including the MOMGA-II. *Proceedings of the First International Conference on Evolutionary Multi-Criterion Optimization,* pp. 226-240, E. Zitzler et al. (eds.) Vol. 1993, Lecture Notes in Computer Science, Springer, Berlin.

Chapter 13

A SURVEY OF PRACTITIONERS
OF EVOLUTIONARY COMPUTATION

Gregory S. Hornby[1] and Tina Yu[2]

[1]*University of California Santa Cruz & NASA Ames Research Center, M/S 269-3, Moffett Field,
CA 94035;* [2]*Department of Computer Science, Memorial University of Newfoundland, St. John's,
NL, Canada*

Abstract To assist in understanding trends in the field of Evolutionary Computation (EC)
and in helping graduates find jobs in EC, we conducted a survey from March 2005
to February 2006 on members of the EC community. The analysis reveals various
technology transfer strategies and activities took place during the past 50 years:
parallel exploration of multiple application areas; a combination of exploitation
and exploration approaches to develop EC applications; and the healthy migration
of EC practitioners between different parts of the globe. We believe these emerged
and self-organized phenomena contribute to the growth of the field. While there
are still challenges in deploying evolutionary computation to industry in a grand
scale, the EC community demonstrates the adaptability and resilience necessary
to achieve that goal.

1.   INTRODUCTION

The field of Evolutionary Computation (EC) has been around for several
decades (De Jong, 2006; Fogel, 1998). In recent years, there has been an ex-
plosion not only in the different types of biologically inspired algorithms, but
also in the number of practitioners in the field. A critical part of this growth and
development of the EC field has been the technology transfer of EC from academia to industry and the successful application of EC techniques to real-world
problems. To assist in the continued technology transfer of EC techniques from
academia to industry, we conducted a survey of EC practitioners working in
both academia and industry. This chapter summarizes some of our findings.

The survey was conducted between March 1, 2005 and February 28, 2006 by
posting 14 survey questions on the SIGEVO web-site. The survey we ran had
three parts. First, it asked several questions about the participant's background.

G.S. Hornby and T. Yu: *A Survey of Practitioners of Evolutionary Computation*, Studies in Computational Intelligence
(SCI) **88**, 283–297 (2008)
www.springerlink.com

The second part of the survey had questions on their job information, and the third part was only for non-academic jobs and asked about EC acceptance and applications at that organization. In this survey, we are particularly interested in learning about how participants found their job, how EC techniques are looked on in their organization, which applications they used EC, what problem types that they found EC to be useful and what obstacles they have encountered in applying EC in their organizations.

In this chapter, we report our findings and our observations of the trends in the EC field. Some of the main findings from our results are: there has been an exponential growth in both EC graduates and practitioners; the main source for finding a job has been networking; while most respondents to our survey are in Europe, the most growth of EC in industry has been in North America; the main application areas of EC techniques are multi-objective optimization, classification, data mining and numerical optimization; and the biggest obstacle for the acceptance of EC techniques in industry is that it is poorly understood. In the rest of this chapter, we present the methodology of this survey in Section 2. Section 3 summarizes survey participants' personal information. In Section 4, we report EC-related jobs. EC positions, problem types and application areas are analyzed in Section 5. Section 6 provides data on computer clusters used for EC applications while Section 7 gives the EC acceptance in industry. Suggestions for future surveys are discussed in Section 8. Finally, Section 9 concludes the chapter.

2.   METHODOLOGY

The respondents to this survey were not randomly selected but were recruited through a variant of the snowball sampling strategy (Vogt, 1999). The recruiting methods include posting the survey announcement to various EC mailing lists (such as EC-Digest and genetic-programming), e-mailing the announcement to attendants of major EC conferences (such as GECCO-05, GPTP-05, EH-05) and advertising the survey at these conferences. Snowball sampling relies on referrals from the initial subjects to generate other subjects. Although snowball sampling may introduce bias into the study, it can be effective in reaching groups having common characteristics (Atkinson and Flint, 2001). In our case, many EC practitioners are likely to subscribe to EC-related mailing lists and attend EC-related conferences, hence they can be reached by our recruiting approach. However, snowball sampling does not qualify as a random process. Consequently, the results from this survey cannot be generalized to the entire EC practitioner population, regardless of the number of responses received. Nevertheless, these results are still useful for gaining a preliminary picture of EC-practitioners in the world.

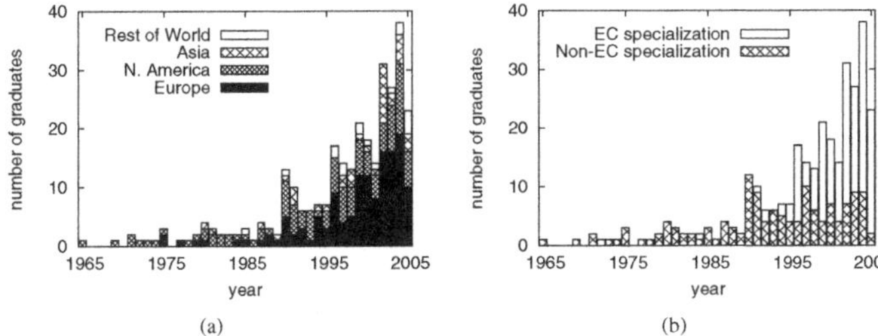

Figure 13-1.   This box-plots in this figure break down graduates by: (a) the geographic region in which they studied and (b) whether or not their degree was specialized in EC.

Over the one year time period in which the survey was taken, 324 responses were received, of which 305 had some EC relation, either through graduating with a degree specialized in EC or by using EC in one of their jobs. For the results of this survey, only the 305 responses which had an EC connection were used.

3.   PARTICIPANTS INFORMATION

The first part of the survey asked participants to provide their personal information, such as gender, and to answer some questions about the most recent degree that they had received. Of the 305 responses with an EC connection we found that 71.1% of participants have a Ph.D. and the gender split is 87.5% male and 12.5% female. Looking into the geographic regions from which participants graduated, we found that most participants graduated from Europe (46.2%), which is followed by North America (35.7%), Asia (12.5%), Oceania (2.6%), South America (2.0%), and Africa (0.7%). A small number (0.3%) of the survey participants did not answer this question. This data is shown in Figure 13.1(a). When we looked for yearly trends in these percentages based on graduation date, we found that they have remained fairly constant throughout the years.

One change that has occurred over the years is in the amount and the specialization of graduates. As shown in Figure 13.1(b), there is an exponential growth of graduation rate, starting with only a couple of people graduating a year from the 1960's up until the end of the 1980's, at which point the numbers increased dramatically and reach a peak of 36 graduates in 2004. For the first few decades, none of those who graduated in this time period had a degree specialized in EC. The first EC grad does not show up until 1991, and then starting in 1996 the majority of graduates have an EC-specialized degree. This suggests that EC emerged as a field of its own sometime in the mid-1990s.

4.   JOB INFORMATION

In total, there were 540 jobs entered by the participants, of which 424 jobs used EC techniques. For determining trends by year, the jobs that were entered by the participants were converted into *yearly positions*. That is, a job from 1997 to 2001 was separated into five positions: one in 1997, one in 1998, and ones in 1999, 2000 and 2001. To limit participants such that they had at most one position in each year, jobs with overlapping years were modified so that the second job started in the year after the first job ended. For example, if a participant had a job from 1995 to 1998 followed by one from 1998 to 2001, the starting year for the second job was changed from 1998 to 1999. Using this method, the 540 jobs were mapped to 3,392 *yearly positions* and the 424 jobs that used EC techniques were mapped to 2,955 EC-related positions.

4.1   Job Sources

After graduation, the next career step for most people is finding a job. By far the most common source of a job was networking, through which 35.1% of participants found a job. This was followed by *other* (25.7%), *supervisor* (17.2%), *postings at university department* (13.8%), *web* (4.7%), *campus career services center* (2.0%) and *mailing list* (0.7%). Looking for differences between those who took a job in academia versus those who took a non-academic job, we found that *networking* was used more for finding a non-academic job (43.0%) than it was for finding one in academia (31.5%). In contrast, the reverse was true for *postings at the university department*: it was used by 16.5% of those who took an academic position but by only 8.0% of those who took a job in industry. Of those who selected *other*, 19 found their position through a listing in a journal or society magazine (such as the Communications of the ACM and IEEE), 13 found their job through an advertisement in the newspaper, 11 founded their own company, and 6 applied and received a research grant.

Further examination of the correlation between the job areas and the job-hunting methods found only a couple of patterns (Figure 13.2(a)). One is that *postings at the university department* helped in finding jobs in Energy, Robotics and Government laboratories, but was of little use for the other job areas. Similarly, the *campus career center* had some success only in finding jobs in Government laboratories and Other. When job-finding methods are analyzed with the job regions, it shows some additional regional trends. *Networking* was used to find over half of the jobs in North America (as well as in Africa, Oceania and South America), but for less than a third of jobs in Europe, and for only 15%

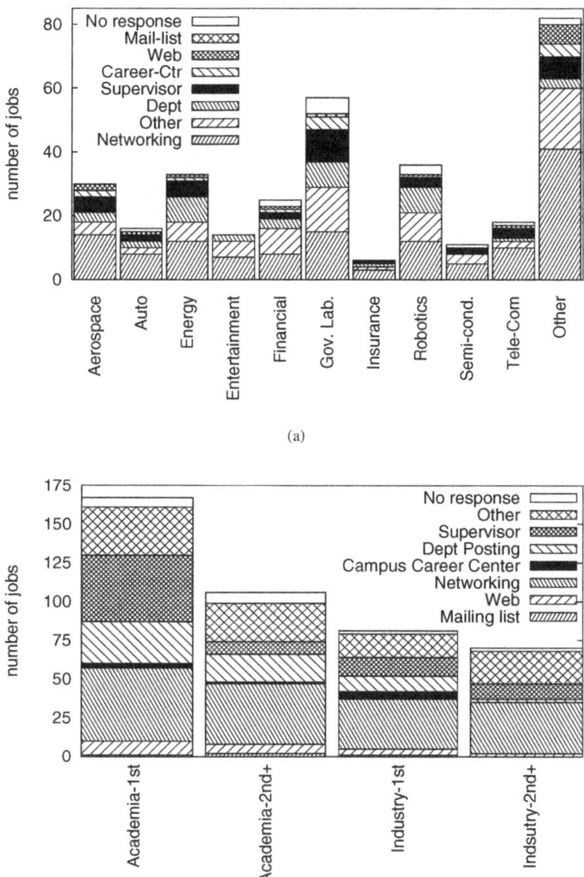

Figure 13-2. This figure contains two graphs which show which job methods were used to find a job by: (a) job area and (b) those going into academia versus those going into industry for their first and later jobs.

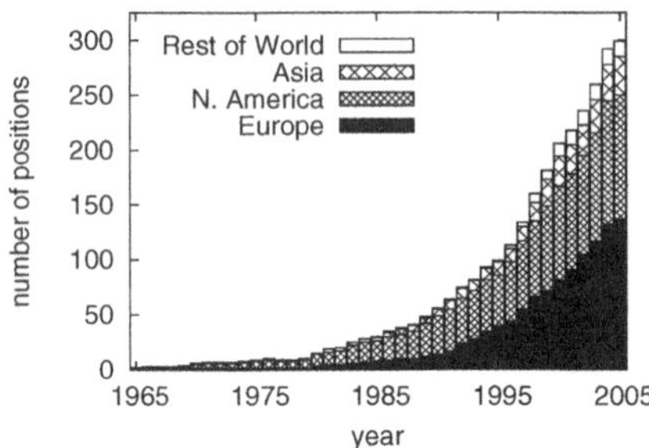

Figure 13-3.   A breakdown of EC positions by year and geographic region.

of jobs in Asia. In Europe, *supervisors* helped to find roughly a quarter of all jobs, and they were also helpful in Asia but were not very useful for finding jobs in North America. The *campus career center* was used by a small percentage of the respondents in Asia and North America, but was not used in any other geographic region.

One useful information for job seekers is that different methods were used to find the first jobs and subsequent jobs, according to the survey participants (Figure 13.2(b)). *Supervisors*, not surprisingly, are used a lot more to find the first academic position than to find a second academic position. In contrast, *supervisors* has similar success in helping finding the first job and subsequent jobs in industry. For both academic jobs and jobs in industry, methods specified in the *Other* category were used more frequently for a second job than for the first one. This difference is about 5% for academic positions and about 11% for industry positions, which indicates that the participants used more creative approaches to find their second and later jobs.

4.2   Job Regions

Looking into the distribution of jobs by geographic region, we found that most EC jobs have been in Europe (45%), followed by North America (37%), Asia (10%), Oceania (3%), South America (2%) and then Africa (2%). This geographic distribution of jobs matches closely to the geographic distribution of graduates and suggests a strong correlation between where a graduate studied and where s/he worked. Also, the ratio of positions between the different geographic regions has been fairly constant over the years (Figure 13-3).

When the job positions are grouped by geographical regions, analyzing the responses over the years reveals that the ratio between positions in industry and in academia has been fairly constant in recent years both in Europe (1:3) and in North America (2:3). In contrast, Asia has experienced a shift in its ratio from being predominantly in industry (100% non-academic in 1981) to being predominantly in academia (more than 75% academic in 2005). For the other geographic regions, the numbers of respondents was too small to give a meaningful interpretation.

Examining the movement of EC graduates for work reveals some interesting trends. First, none of the respondents who graduated with a degree specialized in EC from Africa or South America have left their regions for a job and only 12% of people who graduated in Europe or North America ever move to a different region for work. In contrast, 44% of EC graduates in Asia and 40% of EC graduates in Oceania move at some point after graduation. Second, the direction of movement in Asia, Europe and North America is toward the West. Of those graduates who moved to a different geographic region for a job we found that: 62% of those graduating in Asia moved to Europe at some point, but only 25% ever moved to North America; 70% of those graduating in Europe moved to North America but only 20% moved to Asia; and 67% of those graduating in North America moved to Asia but only 17% moved to Europe for a job. Thirdly, for those people who moved from another region to North America, half moved for jobs in academia and half for jobs in industry, but for those participants who moved to a region other than North America, in all cases they went for academic positions.

5. EC POSITIONS, PROBLEM TYPES AND APPLICATION AREAS

Once in a job, we are interested in what kind of position in their organization the respondent had, as well as whether or not EC was used and how it was applied.

From our responses we found that there has been an exponential growth in positions in the field, starting with a single EC position in 1965 to just under 300 EC positions in 2005. Breaking this down into academic and non-academic positions, there has been a fairly steady proportion of just under a two-thirds of the positions in a given year being academic and just over a third being non-academic. Figure 13.4(a) is a breakdown of the type of position held for those not working in academia. This figure shows that most industrial EC positions are in research, with a significant number in technical/software development and consultancy. Looking into the types of problems that respondents worked on, we found the following: 40.3% do Multi-objective optimization (MOO), 38.4% do Numerical optimization, 38.0% do Classification, 37.7% do Other, 31.6% do Data mining, 21.2% do Open-ended design, 21.2% do Scheduling, 13.9% do Planning, and 10.1% do Satisfiability/TSP. These values do not add

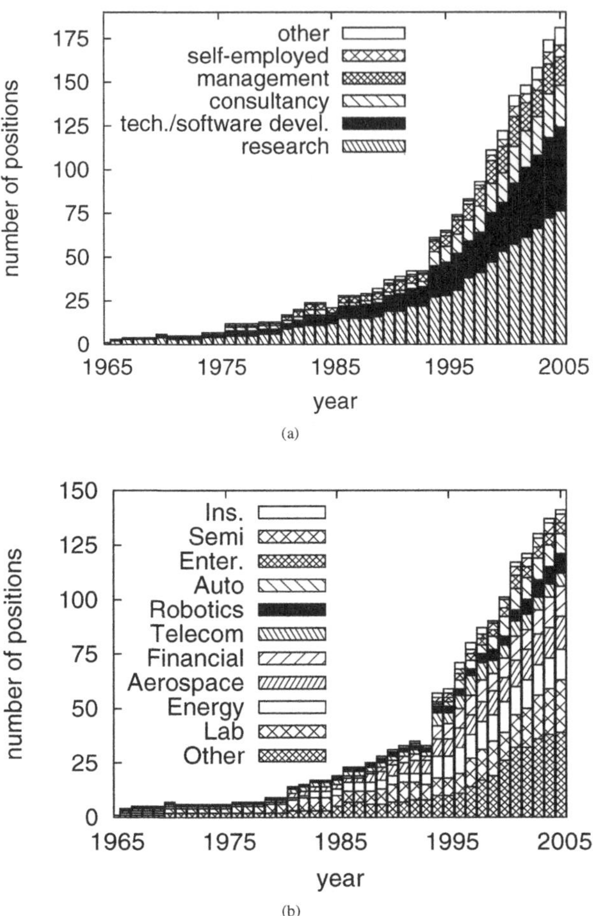

Figure 13-4. This figure contains box plots of: (a) the type of industry position held; and, (b) the application area to which EC is applied.

Table 13-1.   Percentage of respondents working in each problem area.

Area	Percentage (%) working in this area	
	Academia	Industry
MOO	38.8	45.9
Classification	38.0	46.6
Num. opt.	36.7	47.0
Other	39.4	34.4
Data mining	28.7	38.6
Scheduling	19.8	37.6
Open-ended design	24.7	24.2
Planning	13.5	21.3
Sat./TSP	10.1	15.6

up to 100% because participants were able to make multiple selections for each job.

For those responses that selected "Other", participants were able to enter a response in a text field. The most popular entries that were given are: optimization and design (24); modeling and simulation (17), EC theory (15); biology and bio-informatics (11); control (11); evolutionary robotics (6); artificial life (5) and neural networks (5). Many of these entries for "Other" fit under the given categories (eg. 'optimization and design' fits under Optimization and/or Open-ended design) with some of the other entries being an application area and not a problem type.

Comparing the distribution of problem-types worked on by academics to that of non-academics, we found a significant difference (see Table 13-1). In general, the percentage of academic positions that are working in a particular problem area is lower than that for non-academic positions. This means that academics tend to focus on fewer problem areas than those outside of academia. Specifically, those participants employed in academic positions average working on 2.24 problem areas whereas those in non-academic positions average working on 2.74 problem areas. Normalizing for this difference, *Scheduling* stands out as the one problem area which is significantly under-investigated by academics as compared to non-academics.

The distribution of problem areas can be further broken down by examining differences by geographic region (Figure 13-5). First, there are more people working on almost every problem area in North America than there are in Europe, even though there are more EC practitioners in Europe than in North America. Since one position can work on multiple problem areas, this indicates that each position in North America tends to work on more problem areas

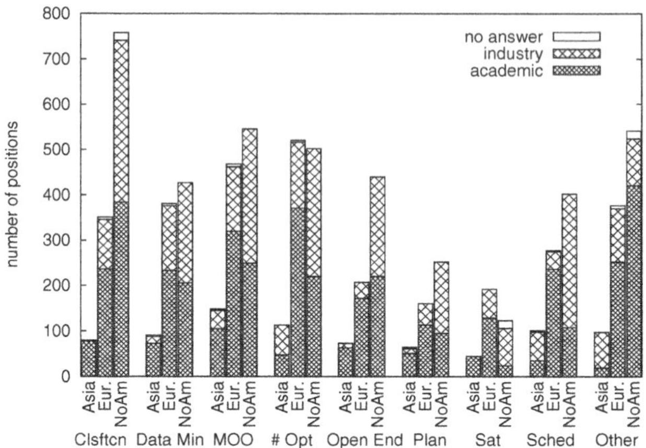

Figure 13-5.   A breakdown of EC positions by geographic region, problem area and academia
or industry.

than a position in Europe does. In fact, the average number of problems areas worked on by each position for the three different geographic regions is
(academic:industry): Asia (2.41:1.92); Europe (2.14:2.39); and North America
(2.35:3.30).

Different from the kind of problem being worked on (numerical optimization,
scheduling, ...), is the industry to which this problem is being applied (automotive, insurance, ...). Figure 13.4(b) contains a histogram of EC industrial
application areas by year.[1] The industry with the largest selection rate is *Other*,
which was selected in 37% of all jobs. The most common areas given by those
who selected *Other* were: IT (13), consulting (12), biology/medicine related
(e.g. Bioinformatics, biomedicine, pharmaceutical) (10), defense and military
(7), and various types of engineering (civil, structural or manufacturing) (7).
For non-academic jobs, the ways in which EC is reported to be most useful are:
design (52.3%), *operations* (33.1%), invention (27.8%), *testing* (15.9%) and
other (14.6%). Of the 31 responses for *other*, 10 were for optimization.

---

[1] Since each job was allowed to enter multiple application areas the total number of selected application areas
can be greater than the number of positions.

Table 13-2.   Percentages of job areas that involve work in different problem type.

Job Area	Tot	Percentage working in this problem type.								
	#	Clsf	DM	MOO	NO	Design	Plan	S/TSP	Sched	Oth
Academic	27	30	33	48	56	19	7	11	15	30
Aerospace	20	50	45	60	55	30	15	20	40	25
Auto	10	40	60	80	80	20	40	20	50	50
Energy	20	70	50	65	50	15	45	10	50	20
Enter.	6	67	33	67	50	67	17	17	50	33
Financial	20	65	80	55	35	20	20	5	35	30
Gov. Lab	34	35	32	41	38	29	15	9	24	29
Insurance	5	100	80	60	20	0	20	0	20	0
Robotics	12	33	33	50	50	25	17	8	33	50
Semi-con	8	50	25	62	25	38	0	12	12	0
Tele-com	12	58	58	50	42	17	17	33	33	17
Other	56	48	38	54	36	20	21	9	32	41

Next we looked into how application area varied by industry to see which combinations stand out (Table 13-2). Some specific combinations that we found are that those working in the automotive and robotics industries are interested in multi-objective and numerical optimization problems, while people working in the energy and entertainment industries are interested in multi-objective and classification problems. Finally, those working in insurance, telecommunications and the financial industries are predominantly interested in classification and data mining.

6.   CLUSTER SIZE

Computer power can impact the applicability of EC on certain applications. We summarize the computer cluster size used by participants in their jobs as follows: 54.2% uses no cluster (single computer), 15.3% uses 3-10 processors, 24.8% uses 11-100 processors, 4.0% uses 101-1000 processors, and 0.7% uses more than 1000 processors. In general, there are no significant differences in cluster size between academics and non-academics. One exception is in the 11-100 processor range, where 27.1% of academics use a cluster of 11-100 processors while only 20.5% of practitioners in industry do.

When computer cluster sizes were analyzed with problem type, Table 13-3 shows no significant pattern. However, when they are compared with application area, Table 13-4 gives some interesting trends. A couple of trends that can be seen are that the automotive and aerospace industries use large clusters of 101-1000 computers more than other application areas do and that the insurance industry uses only a single computer or two. Two respondents use cluster sizes of more than a thousand processors, of which one is a non-academic consultant

Table 13-3.   Percentages of problem types using particular cluster sizes.

Problem type	Total # Jobs	1-2	3-10	11-100	101-1000	1001+
Classification	156	57	16	20	5	0
Data mining	130	53	16	24	5	0
MOO	163	48	19	28	3	0
Num. Opt.	154	47	18	30	3	0
Open Design	88	37	21	34	6	0
Planning	56	46	25	28	0	0
Sat/TSP	41	46	14	34	4	0
Scheduling	88	43	20	32	3	0
Other	143	51	12	26	7	1

Table 13-4.   Percentages of job areas using particular cluster sizes.

Job Area	Total # Jobs	1-2	3-10	11-100	101-1000	1001+
Academic	282	55	12	27	3	0
Aerospace	30	41	20	20	17	0
Auto	16	6	26	40	26	0
Energy	33	51	33	15	0	0
Enter.	14	28	50	14	7	0
Financial	25	60	16	24	0	0
Gov. Lab	57	48	12	32	7	0
Insurance	6	100	0	0	0	0
Robotics	36	30	22	41	5	0
Semi-con	11	63	18	18	0	0
Tele-com	18	61	16	22	0	0
Other	82	50	25	20	2	1

in Europe using it for agent-based simulations and the other is an academic in North America using it for classification and information retrieval.

7.   EC ACCEPTANCE IN INDUSTRY

Next we examined non-academic jobs to see what trends exist in the distribution and acceptance of EC in industry. Even though there is an exponential growth in the number of yearly EC positions, the ratios between the different levels of distribution and acceptance has remained fairly constant throughout the years. The acceptance rate has averaged: 41.3% well accepted; 19.8% accepted; 36.9% somewhat accepted; and 2.0% rejected. The distribution rate has averaged: 36.4% well distributed; 12.3% distributed; 25.3% somewhat distributed; and 26.0% isolated. That these ratios have remained fairly constant over the years does not mean that EC is not becoming more distributed and accepted in non-academic organizations – in fact, the growth in the number of EC positions implies the opposite. What we cannot determine from our data is whether there is an increase in acceptance and distribution within an organization over time, and this is a question for a future survey.

We also analyzed EC acceptance in non-academic organizations by geographic region. The breakdown of acceptance in Asia, Europe and North America is as follows (well accepted, somewhat accepted, not well accepted, rejected): Asia (53%, 7%, 40%, 0%); Europe (41%, 35%, 20%, 4%); and North America (42%, 21%, 34%, 3%); We do not give a breakdown for the other geographic regions due to insufficient responses.

To increase the acceptance and distribution of EC in industry, it is important to understand the obstacles to its uptake. Based on our responses, we found the obstacles to be: *poorly understood* (39.7%), *too ad hoc* (22.5%), *few successful applications to convince management* (21.2%), *commercial tools were unavailable or ineffective* (20.5%), *Other* (18.5%), *no proof of convergence* (14.6%), and *too hard to apply* (13.9%). In some ways, it is encouraging that the main obstacle is that EC is *poorly understood* because as more universities teach EC techniques, these methods should grow in familiarity and thereby gain wider acceptance in industry. Similarly, with a growth in familiarity of EC, companies may be less inclined to find it "ad hoc". The third main obstacle is *the lack of successful applications*, is being addressed through Real-World Applications tracks at EC conferences and with the Human Competitive Competition held at GECCO since 2004. Finally, *lack of useful commercial tools* suggests a possible market niche for those wanting to achieve commercial success with creating EC software. Among the 27 responses for *Other*, the most common obstacles were: lack of experience/familiarity (9), and too slow or does not scale (4).

8.   COMMENTS FOR FUTURE SURVEYS

Having conducted the first survey of practitioners of evolutionary computation we have some thoughts on changes that should be done for future surveys.

First off, to better understand EC education in universities, it would be useful to ask for each degree received the number of courses taken which EC techniques were taught. This would be beneficial for finding out how wide-spread EC techniques are being taught to non-EC specialists and also to find out if EC is being more widely included in the course curriculum. Similarly, it would be useful to query people as to how many EC-specialized conferences they have attended in a given year, or the average number of such conferences they attended a year over the course of each job. This helps us to evaluate if EC-conferences are helpful in disseminating and educating EC technology for practitioners in industry.

Second, in our questions on asking how well accepted/distributed EC is at a particular company, rather than having categories such as "Well accepted" to "Rejected or poorly accepted" for possible answers it would be more useful to ask for a numerical rating from 1 to 5, or 1 to 10 asking for the degree of acceptance. In this case 1 would be "Rejected" and the highest value would be "Well accepted." Such a numerical system would allow for more fine-grained ranking of acceptance and would allow for numerical processing on how acceptance has changed. Also, it would be useful to ask for the level of EC acceptance at the start of a job and the level of EC acceptance at the end of the job (or its current level of acceptance for jobs in which the respondent is still currently employed at). This would allow for analyzing whether there has been an increase in acceptance of EC at individual companies over time. Another question of use would be to ask for the size of the company or organization which the user is working at. It would be interesting to see if there are trends in the size of organization that uses EC, or in its growth in acceptance.

Finally, in addition to canonical evolutionary algorithms (such as genetic algorithms and evolutionary strategies) in recent years various other biologically-inspired computing algorithms, such as ant colony optimization, artificial immune systems and particle swarm optimization, have been developed. It would be useful to add a question asking respondents about which techniques they have used at each of their jobs so as to track their use and also learn what applications they are being used for.

9.   CONCLUSION

Over the years, the use of EC techniques have grown from a few isolated practitioners into a genuine field with a large community. This first survey on EC practitioners has provided us with a preliminary picture of its development in the world. There has been an exponential growth in the number of EC practitioners and EC-specialized graduates, since the first graduates with EC-specialized degrees appearing in the mid 1990's. After graduation, most survey participants found their jobs through networking or from their supervisors. Encouragingly, along with the growth in EC positions has been a growth in acceptance of EC techniques in industry, with the main obstacle to industry acceptance being that the technique is not well understood. EC has been applied to a wide variety of application areas and different problem domains, among which the most common problem areas are multi-objective optimization, classification, and numerical optimization. Although there are still challenges to the continued transfer of Evolutionary Computation to industry, we hope that the results of this survey will help.

Acknowledgments

We would like to thank Gerardo Valencia for putting together the survey website, and also Erik Goodman, David Goldberg, John Koza and Una-May O'Reilly for their support in conducting this survey.

An abridge version of this article was printed in SIGEVOlution, the newsletter of the ACM Special Interset Group on Genetic and Evolutionary Computation, Volume 2 Issue 1, 2007.

References

R. Atkinson and J. Flint. (2001) Accessing hidden and hard to reach populations: Snowball research strategies. *Social Research Update*, 33, 2001.

K. A. De Jong. (2006) *Evolutionary Computation: A Unified Approach*. MIT Press, 2006.

D. B. Fogel, editor. (1998) *Evolutionary Computation: The Fossil Record*. IEEE Press, Piscataway, NJ, 1998.

W. P. Vogt. (1999) *Dictionary of Statistics and Methodology: A Nontechnical Guide for the Social Science*. Sage, London, 1999.

Chapter 14

EVOLUTIONARY COMPUTATION APPLICATIONS: TWELVE LESSONS LEARNED

Lawrence Davis

VGO Associates, Newbury, MA, USA

Abstract   This chapter is a nontechnical discussion of 12 lessons learned over the course of 25 years of applications of evolutionary computation in industry. The lessons bear on project assessment and planning, project management, project interfaces, and managing change.

Keywords: Evolutionary computation application, project management, interface, change management, hybrid systems

1.   INTRODUCTION

The chapter is a more detailed treatment of presentations made in a plenary session of the 2007 Evolutionary Computation and Global Optimization Conference in Poland and in an Evolutionary Computation in Practice session of the 2007 Genetic and Evolutionary Computation Conference in London.

I have worked on more than 100 optimization projects for industrial and governmental clients, and as a result I am frequently asked to comment on the factors that improve the odds of success for such projects. The twelve recommendations below are my response. They are not technical in nature, but they bear directly on a project's likelihood of being used successfully.

In my experience, no more than 10% of successful prototype evolutionary computation systems are deployed to the field. In what follows, I will list and discuss twelve recommendations that, if followed, improve a project's chances of ultimate success.

The comments here are made from the point of view of an external consultant involved in technology transfer into industrial organizations. However, I believe they are just as relevant to technical groups based inside those organizations who are proposing projects using optimization technology.

L. Davis: *Evolutionary Computation Applications: Twelve Lessons Learned*, Studies in Computational Intelligence (SCI) **88**, 299–312 (2008)
www.springerlink.com   © Springer-Verlag Berlin Heidelberg 2008

2.   THE TWELVE LESSONS LEARNED

The twelve recommendations fall into three categories. First there are seven recommendations for project assessment, planning, and staffing. Next are three recommendations for project management. Finally, there are two more abstract and general recommendations for success in industrial optimization projects. The twelve recommendations are described in each of the sub-sections.

2.1   Have a Project Champion

In the terminology of technology transfer, a "champion" is an influential member of the client organization who is passionately in favor of your project.

Champions play several critical roles for an optimization project. They are project advocates inside the client organization who are known and trusted by managers in that organization. They are useful background and domain experts. They can connect the project with the appropriate client employees to serve as assessment and oversight team members. They can translate the technologies being used in the project into the language of the client organization. Finally, when their success is related to the success of the project, they can be highly motivated to help bring the project to success.

A project without a champion may feel like an orphan in the midst of other projects that do have champions and that are competing for resources. A project with a champion has an advocate – someone who can help steer the project through the many phases of the process that will bring it into use.

There are numerous examples of champions who have benefited their projects, and I will cite two here. The first one was a VP of Research for a large agricultural firm. Without his advocacy, a project that was ultimately highly successful would not have made it through the stiff resistance put up by experienced farm managers who did not believe that computers could improve on their practices.

The second was a researcher in a major automobile company, who had studied evolutionary computation and believed that systems using evolutionary computation could produce significant benefits. Without his advocacy, the story we had to tell that company would have been drowned out in the clamor of a number of competing projects with similar claims for their benefits.

2.2   Make the Financial Case with an Assessment

Projects that are characterized as research projects have a much lower probability of being funded and being used in the field than projects that are characterized in economic terms.

What this means for project success is that the client company should be aware from the start that the goal of the project is increased profits, decreased costs, and/or increased levels of service, rather than research, exploration, or learning about the client's business. When the effects of a project are couched in economic terms, it is much easier for a company to decide to support it.

In my experience, the most useful way to begin a project is to carry out an assessment activity lasting from several days to two weeks. An assessment team should include project managers as well as representatives of all the groups in the client organization who will be affected if the system is successfully deployed in the fields. The goal of an assessment is to describe optimization opportunities for the client organization, propose a plan for capturing optimization gains, and quantify the benefits to the organization if the plan is carried out.

It is critical that members of the client company be involved in the planning and estimating parts of this activity, because their validation of the estimated benefits of a project are regarded as more reliable than estimates made by individuals outside the client organization. In my assessments, I prefer to look for projects that will provide ten times as much return per year as their one – time cost. A return of this magnitude means that the project costs will be repaid in months, and will continue for a long time. This level of return is well above the level required for approval in most organizations, and so the primary decision an executive must make in deciding whether to fund such a project is, "Do I believe these numbers?" Having members of the client organization on the assessment team, and having them sign off on the financial estimates, makes it much easier to answer this question.

In my experience as an external consultant, a paid assessment is a much more valuable activity than an assessment carried out at my own expense. It is almost never the case that a client who has been given an assessment at no cost goes on to fund a project. It is almost always the case that a client who has funded an assessment goes on to fund the initial stages of the project. An additional consideration is the fact that a client who has funded an assessment activity tends to give the assessment team more access to critical members of the organization, and tends to take a greater interest in the team's conclusions. I have never had a client who felt that an assessment of this type did not repay the client's investment, although many have felt before the assessments that assessments appeared to be a waste of resources.

Let's consider two examples of successes of the assessment approach, taken from my experience. The first was for a company making mechanical parts that are included in other firms' finished products. The assessment was centered on improving the client's demand forecasting accuracy. The assessment was carried out over two weeks, and involved meeting with company employees at eight locations. The team included three of us from outside the company and three company employees.

The assessment had several important outcomes. It was the first time that the company had an understanding of the company's demand forecasting process from top to bottom, contained in a single person's mind. It was also the first time that many individuals in the company had been asked for their recommendations for improving the company's demand forecasts. Finally, the assessment yielded a solution to the demand forecasting problem that none of us at the beginning of the assessment would have foreseen, and that none of us individually would have been able to design. It was the intense interaction of all the team members that created the ultimate solution to the demand forecasting problems.

The second example was an assessment for a client with interests in data mining. We used the week-long assessment period to gather a list of thirty questions that members of the client company thought would, if answered from the company's data, most impact their profits and work processes. The CEO of the client company said that that one list itself was worth many times the cost of the assessment, and the project continued with data mining to answer the questions that the client company ranked as most important.

2.3   Use the Right Technologies

This recommendation may seem obvious, but it is worth stating anyhow. It is important that an optimization expert be perceived as conversant with, and open to, a wide variety of optimization techniques when considering which techniques to apply to a client's problem. If an expert appears to be conversant with a single technology (evolutionary computation, for example) and seems to be looking for problems that can be solved by that technology, the expert is much less likely to be taken seriously when he or she proposes the use of that technology.

It is worth noting that, although my specialty is evolutionary computation, in the course of more than 100 industrial projects I have used evolutionary computation in the systems we have created less than half the time. Many other optimization techniques, including targeted heuristics, linear programming techniques, ant colony optimizers, simulated annealing, and tabu search, have been the sole optimizer or have been one of multiple optimizers in the systems I have helped to develop. Clients are encouraged when their optimization teams consider a wide range of options, and recommend the appropriate approach for solving their particular problems.

Two examples of this principle come from work with Air Liquide North America, an industrial gas company based in Houston, Texas. I led a team from NuTech Solutions that created an optimization system for Air Liquide's liquid gas production and distribution activities. The liquid gas optimization system included an ant colony optimizer to control distribution, a genetic algorithm to control production, and multiple heuristics to speed up the production planning and seed the algorithm with good solutions.

We also created an optimization system for Air Liquide's pipeline operations. The pipeline optimization system included mixed integer programming techniques, a genetic algorithm, specialized optimization techniques for well-behaved subsections of the pipeline, and optimization heuristics developed by members of the client's Operations Control Center.

The two systems just described are unique, to my knowledge, and each solves problems that had not been solved before. The different optimization techniques listed above each had a critical role to play in the success of these projects.

2.4   Involve the Users of the System at Every Step of the Way

One of the most common reasons for the failure of a successful prototype to become fielded is resistance on the part of its intended users. There are many ways to reduce or forestall such resistance. In my view, the best one is to make sure that the prospective users are involved with the system assessment, design, testing, and implementation at each step of the way.

It can also be important to see that the project is organized so that the ultimate users of the system are incentivized to welcome the project's success. This means that if their bonuses are determined by their meeting specific performance goals, the success of the system is either explicitly made one of those goals, or is recognized by the users as a way to help them meet their goals.

If a system's users are incentivized to achieve certain levels of productivity, and if the development of the system takes time away from achieving productivity, they will not welcome participation in the project. If the users' bonuses are tied to productivity and the system shows promise for dramatically increasing productivity, the users may be enthusiastic about participating in the system's assessment, design and implementation activities.

I am not aware of any project with an explicit goal of replacing humans that ultimately succeeded in being fielded. Under such conditions, the human experts tend to withhold critical information from the project team, criticize the system for doing the wrong thing with regard to that information, and give the system a highly negative rating.

Projects with the goal of offloading repetitive, unpleasant activities from humans have a good chance of success. So do projects that help them to foresee emergencies or help them to do the hard work involved in detailed planning.

One example of the importance of involving the users at each step of the way is the agricultural project mentioned above. The potential users of the system – the farm managers – rejected every overture from the project team, and refused to be involved in the project design, testing, and implementation. The system's eventual fielding took three extra years and required the termination of more than half of the farm managers.

Another example is an optimization system produced for Chevron, in which the oil field operators were initially very resistant to the project, fearing that it would replace them as planning experts. The system was designed so that the operators could edit all information flowing in, could select which optimization techniques would be used, and could edit all schedules produced by the system. It was clear to the operators that they would be in control of the system rather than the other way round, and at present this very successful system is regarded as one of the most impressive in use at that site.

2.5   Don't Speak Technically

This recommendation can be one of the hardest for a practitioner of optimization to adhere to, but it can be critical. In general, it does little good for a technology transfer team to do detailed presentations of the technology they plan to use, and it can do significant harm to the project. Executives and managers are accustomed to buying software packages with algorithms inside that are not described on the box, and many of them prefer it that way.

Pursuing technical descriptions of evolutionary optimization algorithms, or tabu search, or ant colony optimization, or linear programming algorithms can lead to client rejection because such presentations show the client that he or she does not really understand what the team is doing. When it comes time to decide whether to spend money on an optimization project that is somewhat baffling in nature, or to spend the money instead for additional training for a sales team, there is a strong tendency to fund the activity that the manager understands.

Early on in my career, I could be observed saying things to clients on the order of: "An evolutionary algorithm with a permutation representation and uniform crossover is used to solve the problem." Few of those systems ultimately made it to the field. At the present time, the way I would say that same thing goes something like this: "The system creates tens of thousands of solutions, and works hard to make them better and better."

One example of the use of this principle was an assessment at a mid-sized energy company, in which we never described the technology inside the software system we would build, and only described the things that the system would do for the user (things no other system currently does). Another example is the agricultural system mentioned above. For our first two years on that project, we were forbidden by our project champion inside the company to utter the phrases "genetic algorithm" or "optimization" on company property.

2.6   Pay Attention to the Politics

The successful fielding of a great optimization system depends on a large number of factors that are unrelated to its technical components. The most critical is the management of the perception of the system inside the client organization. People outside the project team cannot see or understand the algorithms used by the system. What they can see and understand are the system interface, system demonstrations, and the reactions of other people in the client organization to the system idea.

It is critical to manage the perceptions of other people in the client organization if the system is to be successful. Managing perceptions, however, is not a skill that is typically taught in academia, and is more the province of project champions, managers, and executives.

There is a great deal of diplomatic work to be done in moving a system through the various stages of approval and acceptance in a client organization, and it is critical that a project team have at least one member skilled in such diplomacy.

One example of this principle in action is an oil well work-over project carried out for a major energy company. In the end, the project involved transforming the way that the workover teams thought about and planned their work. Some of the workover specialists had more than thirty years of experience, and were not disposed to change the way that they did things.

The largest challenge for us was to get the specialists to consider an alternate way of doing their work. We used a wide range of techniques to manage their interactions with the system, including weekly project calls to inform them of the project results, compelling graphical displays of the approaches to be used, creation of two champions among them who believed that the new approach would make their lives easier, and recruitment of the influential workover telephone dispatcher who saw that her job would be much easier if the system were to be employed. In addition, there were frequent meetings with upper-level managers to acquaint them with the status of the project and the way that the project would impact the field workover crews.

Interactions with the various groups who would be affected by the project took up a substantial amount of project time, and this was entirely appropriate, given the magnitude of the potential impact of the project on the organization if it were a success.

Another example is the Chevron project mentioned above. The project was managed under a detailed project management structure developed in Chevron called "CPDEP." Each phase of the project had a review board composed of more than ten members, including prospective users. Each phase was scrutinized for potential benefits, costs, and risks. At the end of each project phase the review board met and carefully considered whether to move forward to the next phase of the project.

Under the CPDEP process, the ultimate users of the project and their managers were involved at each step of the way, and this fact accounted for the project's surviving Chevron's merger with Texaco and the high level of personnel change this entailed. The CPDEP process also guaranteed that the concerns of the users would be listened to at each step of the way.

2.7   Plan the Project in Phases so that Each Phase Delivers Value

It is much easier for a client company to approve a project that proceeds in small phases, each of which is intended to deliver value, than it is to approve a large project whose value won't be known until after the project is done.

Breaking a project up into small phases also gives managers the option to cancel or proceed at each stage of payment—a fact that makes them more comfortable in approving a project in the first place.

Making sure that each phase of the project will deliver value may require some ingenuity. For example, how can a complex optimization system deliver value when its parts need to work together in order for complete solutions to be delivered?

Some ways to provide value in phases include the following: create characterizations of best practices as experts are observed and debriefed, so that the experts will have documents to work from and teach to; observe the metaphors that experts are currently using and look for improvements that will help them capture value before the system is completed; document the results of each phase, particularly if work process documents do not already exist in the client company, so that the company has a better understanding of the work processes that will be optimized; use demonstration versions of the system subsections to inspire changes in current expert work practices.

However you do it, try to make sure that each phase of the project provides great value, starting with the assessment. This is a good way to make managers confident that, when they approve later phases of the project, they will receive value for their investment.

Breaking the project into distinct phases also allows managers to see whether the project is being implemented on schedule and budget. Delays and cost overruns are two of the most frequent reasons for lack of satisfaction with ongoing projects. If your project phases are completed on time and on budget, this will increase the level of confidence that you can complete the project as you have predicted.

One example of the use of this principle was the liquid project distribution system for Air Liquide North America, mentioned above. The project had an assessment phase, a design phase, a distribution phase in which routes were planned given the company's current production schedules, a production phase in which production was scheduled given route planning, a six-month phase in which the system's results were monitored and checked by Operations Control Center staff, and a series of phases in which the various geographical regions of Air Liquide's operations were added to the system. Each of these phases provided value. One of the most interesting and unexpected sources of value was discovered where the experts noted that they had learned many new ways to improve their operations by observing the way that the system did its schedules.

Another example was the well workover project for a large energy company. The assessment yielded many options for increasing profits in the workover process, although the initial project implemented only one of those options.

The workover staff were walked through the way that the system would work before it was ever computerized, to make sure that they understood it and accepted the changes to their routines that its use would entail. Each phase of the project produced a more detailed model of the workover operation. After the first module was implemented, schedules were useful but not detailed. Each new module added scheduling accuracy and additional savings to the schedules produced.

2.8　Create a Great Interface

The user interface to an optimization system has little to do with the actual optimization techniques or their performance, and it has everything to do with how the users think about the system. The typical user lacks the ability to determine how well an algorithm optimizes, and may not be impressed with numerical results before they have been thoroughly examined by the experts, but the typical user knows a great deal about seeing and using interfaces. For this reason, nearly everything the user believes and feels about your system will be shaped by the interface that you provide.

If you have a great optimization algorithm and a mediocre user interface, the user will probably think that you have a mediocre system. If you have a mediocre algorithm and a great user interface, the user will probably think that you have a great system. The interface is what the user sees and interacts with, and for most users the interface is the system.

For these reasons, it is important to have adequate resources in the project budget for the system interface. The interface should track the system at each phase of development, and every demonstration of the system should use an interface tailored to the current system version and capabilities.

To repeat: if the user sees a very good interface to an early version of the system, the user will think that the system is coming along well. In fact, the principal problem in managing the user's perception at this point might be the perception that the system is nearly finished. It is important that the user understand how much work is required behind the screen to add to the system's capabilities.

Because interfaces are unrelated to the real meat of an optimization algorithm, many of us evolutionary algorithm practitioners tend to slight the interface and put resources into additional development of the system algorithms. To do this is to endanger the project's success.

One example of the power of great interfaces is the work of George Danner and Howard Park at Industrial Science, a company founded by Danner. Danner and Park use the graphics engine from a computer gaming company under license, making their simulation and optimization projects appear lifelike and impressive to the observer. On multiple occasions I have had the experience of describing a system to a prospective client with slides and have received moderate responses, then watched the prospect become highly excited when seeing the same systems running in real time with their animated, detailed interfaces.

The difference is only in what the client sees, but this is all the difference in the world.

2.9   Publicize the System's Results

It is important that managers, users, and prospective funders know about the development of the system as it proceeds, and that they have a favorable impression of its progress. Frequent demonstrations of system capability are one way to achieve this result. So are testimonials from individuals in the client organization.

A project team may be working intensely and achieving remarkable results, but if the higher-level managers do not see tangible proof of those results, they are likely to believe that the system's progress is what they are seeing – very little. If the managers see frequent demonstrations of progress with compelling interfaces and frequent statements of support from potential users, then they will believe that the project is going well and producing great results. Never forget that a person's view of a project is primarily based on what is observed, not on what the project plan said would be happening.

One project in which this principle was relevant was the CPDEP approach Chevron used to manage our heavy oil field optimization project. At each CPDEP phase, the project review board convened for presentations of system progress, demonstrations of system capability, and reactions from prospective users. An example of this principle in reverse relates to multiple projects I've been associated with in which at some point the higher-level managers stop soliciting progress reports and demonstrations of project capabilities. None of those projects went beyond the prototype phase. Lack of interest in system re-sults is one of the strongest indicators that a project may be failing in its battle for future approval and resources.

2.10 Understand the Work Processes and the System's Effects on Them

A very large contributor to the success of an optimization project – in many cases the largest contributor – is its impact on the way that people did things before the project began.

If the system will cause the existing work processes to be changed, a great many things will happen, and many of them can make future system users nervous. What happens if the system becomes the expert in a process, and the current human expert is asked to work on other tasks? Will that expert's abilities be devalued? Will the expert's status be reduced?

What happens if the system's use requires cooperation from two or more groups that have not cooperated before? Who will lead the cooperation? Whose bonuses will be impacted, and for better or worse? Which group will be helped or hurt by the cooperation? Whose star will rise and whose will fall? These questions have tremendous impact on the future users of the system, and their answers will shape the users' views of the system.

If the project has an assessment stage, and if the assessment team includes potential users of the system and individuals from groups that will be affected by the system's use, then the current work process can be documented, and the system's impact on those work processes can be predicted from the start. However it is achieved, it is critical that the people who will be impacted by the system understand early on what changes to their work processes will be caused by the system. It is also important that they are ready to accept those changes.

In some cases, it can be important to change the organization's compensation structure to reflect the fact that the work processes have changed. Optimization of a corporate compensation plan can be critical to the success of a computer optimization project, and work process considerations should be in the front of the project staff's minds from the start.

Of course, some optimization systems do not change the work processes of their users, and those can be the easiest to implement. Others create dramatic changes in their users' daily routine, and those require a good deal of management.

One example of this principle in action was the change in the well workover process at a large energy company, discussed above. The project caused a change in the metaphor by which the workover specialists scheduled their operations. Corporate policy needed to be modified in order to make the new approach feasible, and the workover specialists' incentive system needed to be modified to be in alignment with the new approach, so that specialists would not be penalized for doing the right thing. These issues were raised from the very beginning with the workover specialists, and were re-addressed at each phase of the project. The specialists needed to understand what life would be like with the new system, and they needed to believe that life would be better for them. Without these outcomes, the project would have had little chance of user acceptance.

Another example is the heavy oil field scheduling project for Chevron discussed above. The operators needed to understand that they would have complete control over the system's inputs, outputs, and operations. They also needed to understand that the system would make routine a process that they did not enjoy and that took about an hour each day – doing a detailed schedule for the field. Finally, they needed to understand that they would be faced with fewer emergencies and instances of substandard production if the system worked as specified, which would mean that they would be leaving the oil field to go home on time much more frequently. Persuading them of each of these things was a major part of managing the users' perceptions of the project.

2.11   Master the Tao of Optimization

This principle is more abstract and difficult to describe than the preceding ones. It's based on Sun Tzu's statement in *The Art Of War* to the effect that the skillful general wins battles without fighting them. In an analogous way, although we see our task as the creation of computerized optimization systems, we can sometimes use our expertise and experience to help client companies improve their operations without producing any computer systems.

Optimizing without computer systems requires an understanding of an organization's work processes, the metaphors its employees use to accomplish their tasks, relevant optimization techniques, and the effects of changes on employees' work processes if the metaphors they use are replaced. Because this principle is less concrete than the preceding ones, it is probably best to proceed directly to illustrations.

One example of the principle in operation is a consulting engagement done by a BiosGroup team for Southwest Airlines. I was not a member of this team, but the results of the project were so good that I am compelled to describe the project here.

The objective of the engagement was to help Southwest Airlines increase its package delivery rate. At the time of the engagement, the airline was passing roughly 30% fewer packages through its overnight delivery service than competing airlines of comparable size.

The team created a simulation of package delivery operations, and experimented with a wide range of changes to the airline's procedures for handling packages. Finally, they found one non-obvious change that would help the airline to achieve the desired level of throughput. The change could be written in a single sentence and, when the airline employees associated with package handling understood the sentence, the predicted increase in flow was realized. Understanding that a metaphor change or a principled work process change in itself can achieve optimization is an important part of industrial optimization projects.

A second example is that of a scheduling project for a large manufacturing client. After studying the client's operations in an assessment project, we proposed to build a computerized optimization system that would increase the throughput of the client's operations by 10%. I let the client project manager know that they could achieve half that increase without using a computer at all – just by changing the heuristics that the schedulers used to produce their schedules. I noted that we could write everything the schedulers needed to know to achieve this improvement on the back of an envelope.

The company decided to finance the full computer system rather than simply changing the work process, but many companies – smaller ones in particular – would have chosen the inexpensive solution for a 5% improvement.

There are many ways to achieve optimization without using computer systems, simply by using the insights we gain as creators of computer systems. The central theme of all these approaches is that if you are optimizing the organization as a whole – its work processes, incentive structures, and employee approaches to doing their jobs – then you are optimizing at the highest level of performance. Another way to put this theme is that the best optimization projects consider the organization as a whole, not a single optimization problem in isolation.

2.12   Love What you Do

The final principle in this list is somewhat abstract as well. The idea is that if you love what you are doing, other members of the project team will note this and may become more favorable to the project.

For more than 25 years I have been creating industrial optimization solutions, and I still get goose bumps when a computer system finds better solutions than we humans have been able to find, using the processes that we believe have produced us humans. Evolutionary computation is fairly powerful stuff, and saving clients millions of dollars with its use can be a thrilling activity.

Don't forget that we are working in an exciting field, and employing the means of optimization at our disposal to assist clients is a very enjoyable thing to do. Examples of this principle can be found all around you. Most of them lie outside the field of optimization. If you enjoy what you are doing, you will do it better and people will be more disposed to use the results of your work. It works in the optimization area as well.

3.   SUMMARY

Designing and creating a successful optimization system can be a challenging and difficult task. This chapter has discussed twelve suggestions for increasing the likelihood that your optimization projects will be well-designed, used, and appealing to the people who decide whether your project receives resources. I wish you all the best in your real-world optimization projects, and hope that some of the suggestions described above will contribute to your success.

Chapter 15

EVOLUTIONARY COMPUTATION AT AMERICAN AIR LIQUIDE

Charles Neely Harper[1] and Lawrence Davis[2]

[1]*Air Liquide Large Industries U.S. LP, Houston, TX 77056 U.S.A;* [2]*VGO Associates. Newbury, MA 01951 U.S.A.*

Abstract   North American Air Liquide has implemented several innovative applications of evolutionary computation and ant colony optimization. This chapter describes Air Liquide's operations and the two major areas in which optimization has significantly impacted Air Liquide: production and distribution of liquid industrial gases, and production and distribution of industrial gases via dual pipelines.

Keywords: ant colony optimizer, genetic algorithm, production optimization, distribution optimization, evolutionary optimization, industrial gas

1.   INTRODUCTION

Air Liquide (Air Liquide, 1902) is the world leader in industrial and medical gases and related services. The Group offers innovative solutions based on constantly enhanced technologies. These solutions, which are consistent with Air Liquide's commitment to sustainable development, help to protect life and enable our customers to manufacture many indispensable everyday products. Founded in 1902, Air Liquide has nearly 36,000 employees and is present in more than 70 countries. Sales in 2005 totaled 10,435 million Euros.

American Air Liquide Holdings, Inc. oversees the North American operations of Air Liquide. Through its subsidiary businesses, American Air Liquide offers industrial gases and related services to a variety of customers including those in refining, natural gas, chemistry, metals, automotive, chemicals, food, pharmaceutical, electronics, specialty and health-care markets.

Our products are primarily oxygen, nitrogen, and hydrogen along with the services and technology involved in delivering these gases. We separate atmospheric air into oxygen, nitrogen and argon through a cryogenic distillation

C. N. Harper and L. Davis: *Evolutionary Computation at American Air Liquide*, Studies in Computational Intelligence (SCI) **88**, 313–317 (2008)
www.springerlink.com

process and we produce hydrogen by cracking natural gas. We distribute our products through several methods: in gaseous form through nearly 2,000 miles of pipelines or in compressed cylinders and in liquid form, by truck transportation from our plants to our customers' tanks and facilities. More than half the cost of creating and distributing oxygen, nitrogen and hydrogen lies in the cost of energy, as natural gas or electricity. Operating air separation and hydrogen plants, cogeneration units and our pipeline is an energy-intensive business.

In 1999 we began to investigate ways to substantially reduce our production and distribution costs and to find "smart" ways to manage our supply chain. We hired BiosGroup, a complexity science company based in Santa Fe, NM, to help us assess the potential for cost reduction. A result of that engagement was the decision to pursue two separate streams of optimization: one related to reducing the cost of producing and distributing liquid oxygen, liquid nitrogen and liquid argon, and one related to reducing the cost of producing, compressing and distributing gases in our pipelines. Even though they are related, these are two very different ways of delivering products, one by truck and one by pipeline.

2.   INITIAL OPTIMIZATION SYSTEMS

In late 2001 BiosGroup developed a Proof of Concept system for a small area of our business that optimized the distribution of oxygen and nitrogen in liquid form by truck from our more than 40 production plants to more than 8,000 customer sites. This system used an ant colony optimizer to determine truck routes and sourcing from our plants. The performance of this system was very impressive, and we realized that there was a good deal of benefit to be gained from extending the system to schedule the production of our liquid products.

In 2002 BiosGroup created a Proof of Concept system for our pipeline operations. This system used a genetic algorithm to decide how to control the pipeline, and it used a mixed integer programming approach to optimize the operations of the plants. While the system did not perform detailed simulation of the costs and performance of our equipment, its results suggested strongly that there were significant savings to be gained if the pipeline optimization system were to be developed into a full-fledged simulator and optimizer.

3.   THE LIQUID GAS SYSTEM

BiosGroup's consulting operations were acquired by NuTech Solutions, Inc. in 2001 (NuTech Solutions, 2001) and the subsequent development of these systems was carried out by NuTech. Some of the BiosGroup project members continued with the projects after the transition. From this point onward, the continued development of both projects was performed by NuTech Solutions.

In the next major phase of the liquid supply chain production and distribution project, we wanted to find the best solution to plan both production and distribution of our products. It did not seem to us that a good off-the-shelf solution existed that could solve the problem of coordinating production and distribution. The major supply chain software systems optimized first production and then distribution and the results seemed to us to be substantially suboptimal. In fact, we acquired one industrial gas company that had created an award-winning production and distribution optimization system based on a large commercial supply chain product, and its performance seemed to be well below what could be achieved.

It was clear to us that the problem of coordinating production and distribution was not one that could be adequately solved by mathematical techniques such as linear programming, because our plant production profiles were not linear, and neither were our contract terms or plant costs for start-up and shutting down. Most importantly, power costs—the dominant costs for us—were not linear, and they changed at fifteen-minute intervals throughout the day in some areas.

The ant colony optimizer that sourced our orders to plants and scheduled deliveries to our 8,000 customers worked well, but it took a long time to run. We asked a NuTech team to study our problem and determine whether it was possible to produce an optimization system that would integrate both production and distribution (something that the commercial systems known to us did not achieve) while finding high-quality solutions in a six-hour computer run (that is the time between updating our databases at midnight and our need for a 6 am schedule for the next day).

The NuTech team has created a system that we believe to be unique and unprecedented. They have built a genetic algorithm to schedule production at our 40 plants producing liquid gases, and they linked the genetic algorithm to the ant colony optimizer in an ingenious way. A top-level optimizer asks the genetic algorithm and the ant colony optimizer to produce production schedules and distribution schedules. It then evaluates the combination of the production schedule and the distribution schedules in order to find out how well they work together. Each optimizer is then given the feedback from their joint result. In this way, the ant colony optimizer and the genetic algorithm adapt in conjunction with each other to generate integrated schedules, even though neither system is explicitly aware of the operations of the other.

A significant insight derived from this system was the observation that, while the ant system operating alone took many thousands of iterations and several hours to come to a solution, it could run three or four iterations per solution produced by the genetic algorithm, so that the time required to run the two systems linked as we have described was under our six-hour limit.

Today we use the liquid gas system to help us schedule the production and distribution of our liquid products. The cost savings and operational efficiencies are substantial. We are saving more than 1.5 million dollars per quarter at one of our plants by utilizing optimization techniques in a demanding and changing environment.

We are currently extending the liquid production and optimization system in multiple ways, and we expect its benefits to increase as these extensions are completed. We believe that the combination of the genetic algorithm and ant colony optimization greatly exceeds the performance of any commercially available approach to our situation, and we would recommend that a company seeking ways to coordinate and improve their production and distribution operations consider a similar solution.

## 4.		THE PIPELINE OPTIMIZER

There are several features of an industrial gas pipeline operation that are different from natural gas and oil pipeline operations. Since most of the pipelines in the world carry natural gas and oil, the off-the-shelf tools for controlling pipelines are not suited to our operations. In addition, they use optimization techniques that sometimes fail to find optimal solutions—or any feasible solution—when operating conditions change dramatically.

On the strength of the pipeline optimizer Proof of Concept, we asked NuTech Solutions to continue to develop the pipeline optimizer project. The goals of the next phase were to produce more detailed solutions to incorporate more realistic hydraulic models of our pipeline operations and models of plant and compressor operations, and to optimize a pipeline with equipment that is not modeled in other pipeline optimization systems (such as devices that can change their functions on command, dramatically altering the hydraulics and topology of our pipelines).

The system that NuTech produced in collaboration with our team greatly exceeded our expectations. The system uses a genetic algorithm at the top level, a deterministic heuristic for analyzing pipeline subsystems and setting pressures within each subsystem, a combination of brute force search and genetic algorithm at the plant level to optimize plant production, and multiple heuristics for modifying solutions based on their performance.

The performance of the system is impressive. The operators in our Operations Control Center have learned a good deal in the process of analyzing the solutions produced by the pipeline optimizer, and have modified the way they think about the pipeline and respond to mechanical upsets and breakdowns as a result of studying solutions produced by the optimizer. The optimizer's results have substantially lowered operating costs for the pipeline and have helped us plan for the configuration and installation of new equipment to improve the efficiency of our operations. We are continuing to think about improvements to the existing tools.

For proprietary reasons, we cannot state the full impact of the pipeline optimization system. But we can say that given the outstanding performance of the NuTech team, we were very proud to recognize them by flying them to our Houston offices from Poland, Massachusetts, North Carolina, and California for a two-day event and recognition reception.

5.   CONCLUSIONS

The two systems described here have transformed the way that Air Liquide Large Industries U.S. LP does business. We have lowered our costs, improved our efficiency, and increased our planning ability. In one of the media releases jointly issued by Air Liquide and NuTech describing the effects of these systems, Charles Harper said "Our partners at NuTech Solutions painted the yellow brick road for us, they showed us Oz, and then guided us through the journey. We recommend to other companies with similar problems that they too embark on this journey—it has given us a new understanding of what is possible using contemporary approaches to optimization."

Acknowledgments

This article was originally printed in SIGEVOlution, the newsletter of the ACM Special Interest Group on Genetic and Evolutionary Computation, Volume 1 Issue 1, 2006.

References

Air Liquide (1902) www.airliquide.com
NuTech Solutions (2001) www.nutechsolutions.com

Index

9 783540 757702